Telecommunications Cost Management

Telecommunications Cost Management

Brian DiMarsico
Thomas Phelps IV
and
William A. Yarberry, Jr.

CRC Press
Taylor & Francis Group
Boca Raton London New York

CRC Press is an imprint of the
Taylor & Francis Group, an **informa** business

AN AUERBACH BOOK

CRC Press
Taylor & Francis Group
6000 Broken Sound Parkway NW, Suite 300
Boca Raton, FL 33487-2742

First issued in hardback 2017

ISBN-13: 978-0-8493-1101-7 (pbk)
ISBN-13: 978-1-1384-7243-3 (hbk)

This book contains information obtained from authentic and highly regarded sources. While all reasonable efforts have been made to publish reliable data and information, neither the author[s] nor the publisher can accept any legal responsibility or liability for any errors or omissions that may be made. The publishers wish to make clear that any views or opinions expressed in this book by individual editors, authors or contributors are personal to them and do not necessarily reflect the views/opinions of the publishers. The information or guidance contained in this book is intended for use by medical, scientific or health-care professionals and is provided strictly as a supplement to the medical or other professional's own judgement, their knowledge of the patient's medical history, relevant manufacturer's instructions and the appropriate best practice guidelines. Because of the rapid advances in medical science, any information or advice on dosages, procedures or diagnoses should be independently verified. The reader is strongly urged to consult the relevant national drug formulary and the drug companies' and device or material manufacturers' printed instructions, and their websites, before administering or utilizing any of the drugs, devices or materials mentioned in this book. This book does not indicate whether a particular treatment is appropriate or suitable for a particular individual. Ultimately it is the sole responsibility of the medical professional to make his or her own professional judgements, so as to advise and treat patients appropriately. The authors and publishers have also attempted to trace the copyright holders of all material reproduced in this publication and apologize to copyright holders if permission to publish in this form has not been obtained. If any copyright material has not been acknowledged please write and let us know so we may rectify in any future reprint.

**Visit the Taylor & Francis Web site at
http://www.taylorandfrancis.com**

**and the CRC Press Web site at
http://www.crcpress.com**

Library of Congress Cataloging-in-Publication Data

DiMarsico, Brian.
 Telecommunications cost management / Brian DiMarsico, Thomas Phelps IV, William
A. Yarberry, Jr.
  p. cm.
 Includes bibliographical references and index.
 ISBN 0-8493-1101-2 (alk. paper)
  1. Telecommunication--Cost control. I. Phelps, Thomas, IV. II. Yarberry, William. III.
Title.

TK5102.5 .D457 2002
384'.068'1--dc21
                                                                    2002071716

Dedication

To my loving wife, Sheila, for being patient, understanding, and always supportive of what I do for a living.

Brian

To Seung-Jin and Martin, for their loving support.

Thomas

To Carol, Will, Libby, and my parents.
Thanks for your understanding and patience.

Bill

Contents

Acknowledgments

Many individuals have contributed both directly and indirectly to this book. We are delighted to continue our association with the editor, Christian Kirkpatrick, who has provided advice and encouragement all along the way. Special thanks to Kathy Dodds, partner in the Global Risk Management Services practice of PricewaterhouseCoopers. Kathy provided many helpful content and editorial insights. Three other PricewaterhouseCoopers partners, Raymond Slocumb, Dana McIlwain, and Rik Boren, provided the initial support and encouragement that helped us stay on course. Bill Moore, a PricewaterhouseCoopers senior manager on temporary assignment in Australia, gave us valuable "long-distance" advice and recommended vendors that provided additional insights. Other contributors include independent telecom consultant Frank Marino, Jeff Richards of QuantumShift, Lee Miller of Sullexis Systems, and Janie Mendez of Compaq Corporation. Finally, we acknowledge the essential support of our families during the months of researching and writing needed to complete the book.

Brian DiMarsico
Florham Park, New Jersey

Thomas Phelps IV
Los Angeles, California

William A. Yarberry, Jr.
Houston, Texas

Introduction

Telecommunications Cost Management is intended for telecom managers, CIOs, controllers, and others who have an interest in reducing the cost of telecommunications services, equipment, and software. Telecom books on the market today focus on either traditional telecom billing audits or pure technology, with the cost-savings ideas buried deep in the text. In contrast, *Telecommunications Cost Management* provides a blueprint for cost reduction across all the major technologies — from Frame Relay to IP telephony, from provisioning to contract recommendations. Busy decision makers need the specifics quickly, without plowing through details that may be important but do not affect the economics of a project. Traditional bill auditing texts provide recommendations but only within the context of the existing architecture — for example, they might highlight techniques for reducing circuit switched leased lines but omit the pros and cons of leased lines versus alternatives such as VPN or fixed wireless. *Telecommunications Cost Management* presents the key facts up front, with sample calculations for broadband, local access, equipment, and services alternatives.

To give this guide to telecommunications cost management some order and coherence, we include the minimum necessary level of technical explanation, with business-focused pros and cons for each technology. We also add a chapter on traditional telecom billing auditing because the "green eyeshade" approach does indeed result in lower costs, realized as either refunds or process improvements over time.

Based on our experiences with companies and their business challenges, we present the following broad approaches to telecom cost management:

- Managing end-to-end telecom procurement, monitoring, bill payment, and follow-up
- Understanding technology alternatives and what effect they will have on long-term costs
- Reviewing outsourcing options from carriers and other telecom providers

Most organizations can save between 10 and 35 percent of their telecommunications costs by implementing the appropriate steps recommended in this book. Of course, we feel "consultant angst" when making such a bold statement, but generally there are substantial dollars lying on the table and the "cash faucet" is running. For example, one of our clients had been charged for years for some toll-free numbers that, in fact, belonged to another firm. Another client received a large tax refund from the IRS based on a court-upheld clause in the Federal Excise Tax Code. Even a small nonprofit organization was able to reduce expenses by replacing a card in its telephone system, thereby reducing recurring trunking expenses. There are hundreds of such examples. Our intent here is to condense and organize the technologies, tools, and options needed to manage telecom costs. Readers accustomed to traditional telecommunications books will be relieved to know that we do *not* cover the seven-layer ISO stack.

If you want to reduce your organization's short- and long-term telecommunications costs while maintaining or improving service levels, this book will provide a wealth of ideas and resources.

Reader comments and questions are most welcome. Send them via e-mail to billyarberry@bigfoot.com.

Chapter 1

Introduction to Telecommunications Cost Management

Nothing comes amiss, so money comes withal.

— William Shakespeare, *Taming of the Shrew*

Telecommunications is the second highest nonoperating expense for the average Fortune 1000 firm. Most organizations can reduce these expenses by three to fifteen percent; some can cut costs by 30 to 40 percent. The key to achieving and maintaining lower telecom expenses is to understand industry drivers, technical alternatives, and effective telecom procurement and processing techniques.

Many organizations pay above-market prices, buy too much capacity, do not detect billing errors, and use less-than-optimal technologies. Compounding the problem is the extreme conservatism of most internal telecommunications organizations — there is only modest reward for cost management but extreme punishment for any service interruptions. Hence, the "if it ain't broke, don't fix it" mind set. Employees expect dial tone virtually 100 percent of the time and are intolerant of any changes that risk downtime.

The chapters that follow outline technologies and techniques that, if applied with management support, can certainly reduce expenses. Large firms today often operate with minimum analytical staff and do not generally look at costs in nonoperating areas (the "plumbing") as carefully as one would expect. While it may be theoretically possible for an organization to be without telecom waste, the authors have never seen a single example. Like fishing at a trout farm, the potential to reel in significant savings is extremely high.

How to Reduce Telecom Expenses (The "Cliff Notes" Version)

At the most general level, there are only a limited number of ways to reduce telecommunications costs:

- Reduce usage (make fewer calls, use fewer trunks, etc.)
- Outsource telecommunications management (cost savings occur only if this is done properly and will not apply to every organization)
- Find less-expensive suppliers
- Restructure contracts/agreements with existing suppliers
- Monitor and correct errors (a Windfall Associates report shows that billing errors occur in approximately 45 percent of all bills, and generally the errors are in the carrier's favor)
- Use more efficient, less-expensive technology
- Decrease tax payments
- Use more efficient internal processes
- Increase security (prevent losses through toll fraud, for example)

The general steps above are influenced by many trends in industry and the workforce. Examples include:

- Increasing importance of telecommunications in general. Reliance on communications for services continues to increase rapidly. From telecommuting to Web-based procurement, the importance of electronic communications continues to monotonically increase.
- Continuing penetration of the Internet as a dominant force in the telecommunications industry.
- Proliferation of dozens of new technologies, including wireless services.
- Increasing levels of technical standardization counterbalanced by high levels of complexity in the telecom architecture (at the provider and customer level).
- Change in the marketplace from supply-driven ("build it and they will come") to a more conservative market-driven environment ("if you are willing to buy it, we will build it").
- Coexistence of old technologies (copper connecting the customer at the last mile) with many new ones.
- Old technologies that work will remain in the telecom infrastructure for decades.
- Continuing maturity of the outsourcing model.

The last bullet, outsourcing, deserves special treatment. As of this writing a number of firms, such as QuantumShift and ProfitLine, offer comprehensive management of telecommunications functions. The client hopes to receive lower prices and avoid devoting management effort to non-core activities. The outsource provider, sitting between the carrier and the consumer, consolidates resources over multiple clients and earns appropriate management fees for the services. Many of these services directly affect expense management. Examples include:

- Bill payment and auditing
- Consumption report generation (including chargeback)
- Implementation of telecom projects
- RFP development
- Procurement, monitoring, and disconnect of services from carriers
- Contract negotiation on behalf of the client (or directly supply services to client as a reseller/aggregator)
- Network implementation

Exhibit 1, adapted from a business plan developed by Hala Fadel and Sunanda Narayanan at the MIT Sloan School of Management, shows features and benefits that could potentially be provided by a telecom outsource firm.

Outsourcing is not a panacea. If agreements are improperly structured, the savings may not accrue. Also, some organizations may have highly effective internal resources that can achieve the same result without the "middleman." The decision to outsource should be reviewed carefully.

Why Telecom Costs Are So Difficult to Manage

Exhibit 1 hints at some of the industry problems that plague telecommunications services. Following is a generic list of cost management issues faced by most organizations:

- Telecom bills are large (delivered in large boxes or multiple CDs), difficult to read, and often not electronic.
- Telecom vendors (local, long distance, etc.) do not have uniform formats for billing information.
- Correlating consumption (number of minutes used, etc.) to the bill is often difficult.
- Forecasting the organization's future usage is difficult. Trunks and other services must often be ordered in advance, based on an estimate of future need.
- Internal expertise, especially for the newest available telecom offerings, may be lacking.
- Fear of change hampers some initiatives that, if implemented, could reduce expenses.
- Telecom regulations, while simpler than in the past, are still complex (certainly for the United States and increasingly for the rest of the world). For example, some organizations, such as airlines, are exempt from the U.S. Federal Excise Tax for telecommunications.
- Voice, data, and video integration continue. The billing infrastructure for these three media has traditionally been different (fixed months versus per minute, etc.). As some per-minute costs get merged into packet-based, flat-fee services, confusion over billing will undoubtedly surface.
- The telecommunications environment is dynamic. Technologies, carriers, offerings, and pricing changes are almost constant. A study done in 2000 may not apply in 2002.

Exhibit 1.   Potential Outsource Provider Benefits

Client Issue/Need	Products/Features from Outsourcer	Value to Client
Maverick spending on telecom services/ equipment	Expense/consumption tracking at all levels; centralized control over orders	Savings from cost tracking and management at every level
No easy bill-back to clients' customers (e.g., consulting)	Cost tracking per client at employee level	Easy bill-back process; savings from accurate bill-back procedure
No centralized source for telecom resources or information	Resource library with telecom contacts, news, information	Time savings from reduced search time for information
Inappropriate solutions/ services for given consumption patterns	Web-based monitoring of solutions; automatic RFQ generation	Customized/current solution at best price; comprehensive dynamic market information
Little visibility into spending; impossible to compare bills across carriers, services	Bill aggregation across locations, departments, carriers, and services	Clarity on spending and consumption; aggregated volume for good negotiating
Voluminous, incomprehensible telecom bills	Customized billing reports; easy-to-use software interface	Clear, relevant reports for corporate decisions; daily access to handy telemanagement tools
Complex pricing contracts	Contract integration into system	Clarity into contract terms and implications for re-negotiation
High perceived risk in trying new carriers or services	Rating and references	Minimizes risk trying new providers/services; provides a credible comparison platform
Overstaffed telecom management departments	Most comprehensive, online telecom management tool	Savings from reducing telecom-related human resources; client can focus on core business

Drivers for Customer Demands

Many software applications today require increased bandwidth (e.g., distance learning, telemedicine, interactive video, videoconferencing). Accelerating intranet usage and mobility requirements drive bandwidth needs. Some typical questions raised by management include:

- How are telecom budgets developed, and are they optimal?
- What voice and data circuits are in place, how are they used, and what do they cost?

Exhibit 2.   Functional Areas of Concern by "C" Level Executives

Area of Concern	Primary Organization
Telecom cost management	CFO/CIO
Call center (contact center)	CEO/Business units
Service providers	CIO/Facility group (voice services)
Operations	CIO/Facility group
Emerging technologies	CIO
Network convergence	CIO
Customized networks	CIO/Business units

■ How can recurring circuit costs be reduced? What processes and tools will sustain the reduction while maintaining adequate service levels? Specifically, how are the following best controlled?
  – Circuits billed for but not used
  – Circuits with disconnect ordered but still billed for
  – Circuits underutilized and billed for
  – Circuit consolidation/restructuring
  – Architectural changes
■ How can good ("best") practices be best implemented?
■ How can the network be structured to best match current and future business needs?

Exhibit 2 summarizes telecommunications concerns by "C" level executives. Specific questions from these individuals might include:

■ Telecom cost management:
  – Is $10M being spent for technology that should only cost $6M?
  – Are vendor contracts competitive?
  – Are supplier invoices correct?
  – Is there overspending on voice and data services?
■ Call center:
  – Does the call center provide the company with a competitive edge?
  – How good is the quality of service in the call center?
  – How well have the new technologies been integrated into the "contact center" (e.g., migration to e-mail and Web interaction)?
■ Service providers:
  – What circuits/services are there, how are they used, and how much do they cost?
  – Is this the best deal for the business?
  – Is it best to stay with current service providers or to shop around?
■ Operations:
  – What is the risk of operational failure?
  – Are changes in the business environment affecting operations?
  – Are the right people, processes, and technologies in place?
  – Is outsourcing a consideration?

- Emerging technologies:
  - Are there new technologies that improve service levels or reduce costs?
  - Is the technology scalable, and is it flexible enough to adapt to changes in the business?
  - Are new technologies important?
- Network convergence:
  - Is the network design optimized for changing business needs?
  - When will capacity be reached on the network? Will bottlenecks appear suddenly, hampering business activities?
  - Are converged networks a consideration?
- Customized networks:
  - Can the network be tailored to improve service to the users/customers?
  - What hardware/software is available to help?
  - Will business plans force a change in direction?
  - Should any of the following tools be considered?
    - Workflow processing for the provisioning processes
    - Automatic verification of network services inventory and pricing
    - A complete end-to-end process that ensures revenue and expense stream matching and reconciliation
    - An organization focused on value-added analysis rather than score-keeping
    - Electronic invoice feeds from suppliers for network services
    - A central repository for all network service disputes and payments
    - Integrated network inventory management, accounts payable, billing, and planning systems
    - Visibility to all network costs within the organization

Carrier Challenges

Although the focus of the book is cost reduction for the telecommunications consumer rather than revenue enhancement for the provider, it is important to understand the risks and issues faced by the carriers. Some of the more important issues include:

- IP packet services deployment and operational support
- Network convergence (voice, data, and video)
- Competition and deregulation in the local loop ("the last mile," e.g., wireless/cable)
- Advent of the "super" or "mega" carrier
- Infrastructure rebuilds to support new broadband access technologies (e.g., xDSL, satellite)
- Carrier cost containment efforts
- An antiquated revenue model, characterized by most of the cash coming from voice (slow growth) while much of the buildout (cost to carrier) is for data

From the carrier's perspective, emerging technologies have not only increased the number of options available, but are also creating customer confusion (e.g., xDSL, VoIP, VoDSL, VPN). At the same time, the large carriers have started to divest certain business units such as wireless, broadband, and long-distance services.

Carriers must address at least four key drivers of change to enable a successful transition to take place:

1. Customer demand (what services are customers willing to pay for?)
2. Technology
3. Capital markets (available funds allow migration away from legacy systems)
4. Government policies that encourage competition

Also thrown into the mix are segmentation changes and market dynamics. For example, now carriers are competing as niche players in areas such as:

- Internet service provisioning (consolidation is predicted)
- Infrastructure providers (laying of fiber e.g., Global Crossing, Level 3)
- Fixed wireless and cable to bypass local loop
- Global providers (AT&T, UUNET, Equant, MCI, and certain RBOCS), CLECs (e.g., MFS, Teleport)

The delivery or network infrastructure of the local access market is currently segmented into four provider areas:

1. Customer premises provider (requires data access to Internet and other businesses, voice access to LECs and IXCs)
2. Access provider (delivery of services: CLECs, ILECs, integrated communications provider [ICP] that can own, install, and manage CPE)
3. Network provider (delivery of services: connectivity between COs and POPs, ATM, FR, DS-3s, Fiber Rings-SONET)
4. Service provider ("manufacturers" of services such as Layer 2 and Layer 3 VPNs, FR, ATM, TLS, Web hosting from ISPs, LECs, IXCs)

Another competitive pressure item is Multi-Protocol Label Switching (MPLS — routing tags on IP header data). As of this writing, the MPLS standard is nearing approval. When fully implemented, MPLS will allow non-ATM networks to have a measure of QoS (quality of service) and will minimize bandwidth. MPLS networks will be less expensive than frame relay networks of the same bandwidth. All these new technologies present the same questions to carriers: Is this a good bet? Will it pay off quickly enough to justify the capital expenditure?

Many telcos have moved from transport-only providers to providers of data services (e.g., Frame Relay, ATM). The current telco environment is focused on VPN shared services; application-specific services are viewed as a strong future source of revenue. For example, telcos are looking to:

- Support ASPs (application services providers) with QoS guarantees
- Provide Internet broadcast capabilities (audio and video, B2C and B2B)
- Provide VoIP (Voice-over-Internet Protocol) products/services
- Bring together OSS (operations support system) provisioning information — primarily data collected from the switches, including unified billing data and service usage information

Summary

Telecommunications costs can be reduced in virtually every organization. Although changes in technology, markets, carrier issues, demand, and employee requirements seem to escalate continually, there are standard solution sets that can help. By understanding the tools — whether technical, procedural, negotiations, or simply "throwing it over the fence" via outsourcing — the organization can make an informed, best-fit decision.

Chapter 2

Vision

Eighty percent of success is showing up.

— Woody Allen

But the bravest are surely those who have the clearest vision of what is before them, glory and danger alike, and yet notwithstanding go out to meet it.

— Thucydides, 5th Century BC

Success requires vision. This chapter discusses some of the functions that must be considered when developing a cost management vision. It also includes some preliminary fact-gathering recommendations to help estimate the boundaries of potential cost savings.

Develop a Cost Management Vision

Why develop a cost management vision? Controlling costs within the telecommunications network is hard work. Some projects are easy with a high payoff; others require constant review. Without a clear vision of the outcome, good intentions may falter. Similar to affirmations in self-help books, a cost management vision might read as follows:

- The architecture of the network (data, voice, hardware, leased lines, etc.) will, by its structure, minimize costs.
- Monitoring systems will alert management of financial exceptions.
- Configurations and clusters of technology will be flexible and scalable to reflect declining unit costs resulting from technology improvements.

- Network investments will match the business culture — no long-term investments for an organization that demands a very quick payback from all its other capital expenditures.
- Financial commitments for telecom services are flexible and can accommodate acquisitions, divestitures, major application changes, and rapid growth.
- Telecom should be perceived as an asset — not just a commodity.
- All alternatives that best support the core business, including outsourcing, will be periodically reviewed for applicability.

After the telecom cost management vision is developed and tailored to the culture of the organization (risk tolerance, scope and level of desired savings), the next step is to consider how to start.

How Do Organizations "Make It Happen"?

Telecom cost management is both a project and a process. A project is needed to gather information and make the right decisions. The process ensures that any gains will be maintained. When considering how to start the process, management should start with the following questions:

- Does staff exist within the organization to perform the analysis required to manage costs? For large organizations, a telecom cost management project may require months of work by skilled analysts.
- Does the firm have the appetite to consider telecommunications changes (either technical or procedural)?
- Is there a bias for or against outsourcing? One note of caution: even if the business culture is strongly pro-outsourcing, it is important to have at least a high-level, in-house understanding of the telecom environment to ensure that agreements with the vendor of choice are equitable for both parties. Another approach is to use the services of a third-party outsource consultant.
- Do the individuals assigned to perform the initial analysis have the right background for the project? Do they know telecom billing, tariffs, telecom taxation, and industry trends (business and technical)?
- What outside resources are contemplated — consultants, telecom auditing firms, outsourcing firms, or others?

When considering how to move forward with the project, risks should be explicitly considered. For example:

- What would be the effect of changing carriers? Telecom managers generally dread carrier changes, even if they are dissatisfied with the carrier. Changing circuits and other infrastructure often causes some disruption that users notice.
- Are users willing to accept technical changes if they cannot directly see the benefits? For example, consider the change from remote dial-up (using a remote access server) to using a VPN (virtual private network) for

connecting to the network. Until all the ISPs across the country get their account numbers correctly loaded, remote users might occasionally fail to get on the network. In the long run, it is certainly the most economical practice for large numbers of remote workers, but there is some short-term pain in the transition.

■ Are negotiators for contract changes experienced in telecommunications? Strong negotiators can push telecom vendors for rates so low that the result is not a win–win situation. The telco may be tempted to devote attention to other customers. Also, negotiating for the right prices and services is critical. Why negotiate a 5-cent-per-minute rate to the United Kingdom when the firm makes only a small number of calls there each month?

Looking for the Quick Fix

Organizations have many agendas and priorities. Sometimes, telecom decisions are not made for the long run because there are more pressing issues. Like the Russians in World War II who, in desperation, sometimes sent unpainted tanks to the front in winter, business managers have to survive the present and not worry about the rust of the future. Accordingly, there may be times when a quick fix is necessary. Save some money now and go after the deeper savings later.

Following are some considerations and approaches for telecom short-term relief:

■ *Use contingency-based auditing firms.* Relying on splitting the proceeds of finding errors and overbillings, contingency firms become speedy and efficient. Their goal is to send in highly trained, "drill-down" staff; find the gold nuggets; and move on. The downside to this approach is that changes in processes that would prevent the errors from occurring in the first place are sometimes not addressed. In addition, this style of telecom auditing emphasizes reviews of bills, agreements, etc., rather than technology alternatives. The question "Was there an erroneous bill for a T1 after office X closed?" might be asked. The question "Should frame over DSL be used in place of a T1?" will likely not be asked.

■ *Renegotiate the contract for immediate relief.* Many carriers, anxious to lock in a customer for several years, will lower unit costs in return for longer contracts.

■ *Throw telecom "over the fence" to an outsource firm.* In fairness, this may be a perfectly acceptable long-term solution as well. However, if the deal is done quickly and without adequate knowledge on the part of both parties, it might not be optimal for the long term. But certainly if telecom is "out of control" and expense management has not been a priority, outsourcing can likely assuage the financial worries of management (at least for telecom).

It is important to recognize that the above comments are generalizations. For example, Houston-based Teligistics performs both contingency work and some process work, such as long-term "pre-audits" of bills. In other words,

Making cents **of your** telecom **DOLLARS**

tel·gistics

Company:	ABC Company
Contact:	Robert Smith
Address:	1234 Anywhere Street
Phone:	(123) 456-7890
Fax:	(123) 456-7890
Service:	Long Distance

BILL CHECKER™
AUDIT / VARIANCE

Billing Period:	5/1/2001 - 5/31/2001
Consultant:	Kim Daughtry
Direct Line:	(425) 397-8555
Fax Line:	(206) 693-2170
E-mail Address:	kim@teligistics.com
Analyst:	Kim Daughtry

ABC Company - Details	Current Bill				Carrier Contract Rates		Variance		
Services	Volume	Amount	Rate	Unit	Amount	Rate	This Month	%	Per Year
Interstate	85,086.20	$4,169.22	$0.0490	/min	$2,765.30	$0.0325	$1,403.92	34%	$16,847.02
Interstate 800	17,421.70	$853.44	$0.0490	/min	$566.21	$0.0325	$287.23	34%	$3,446.82
Interstate CC	431.20	$86.24	$0.2000	/min	$77.62	$0.1800	$8.62	10%	$103.49
Conference Calls	0.00	$0.00	$0.0000	/min	$0.00	$0.1400	$0.00	0%	$0.00
Alaska 800	0.00	$0.00	$0.0000	/min	$0.00	$0.0325	$0.00	0%	$0.00
Hawaii 800	0.00	$0.00	$0.0000	/min	$0.00	$0.0325	$0.00	0%	$0.00
Canada 800	19.80	$3.96	$0.2000	/min	$1.29	$0.0650	$2.67	68%	$32.08
IntraState	9,224.60	$599.60	$0.0650	/min	$322.86	$0.0350	$276.74	46%	$3,320.87
IntraLATA	2,001.31	$118.08	$0.0590	/min	$70.05	$0.0350	$48.03	41%	$576.41
IntraState 800	4,560.70	$269.08	$0.0590	/min	$159.62	$0.0350	$109.46	41%	$1,313.47
IntraLATA 800	1,100.20	$64.91	$0.0590	/min	$38.51	$0.0350	$26.40	41%	$316.84
IntraState CC	67.30	$14.81	$0.2201	/min	$12.11	$0.1800	$2.70	18%	$32.35
IntraLATA CC	0.00	$0.00	$0.0000	/min	$0.00	$0.1800	$0.00	0%	$0.00
International	597.40	$97.12	$0.1626	/min	$73.31	$0.1227	$23.81	25%	$285.70
PICC	0.00	$0.00	$0.0000	/line	$0.00	$0.0000	$0.00	0%	$0.00
Dedicated Loop	0.00	$0.00	$0.0000	/loop	$0.00	$228.9000	$0.00	0%	$0.00
Primary 800 Line	1.00	$3.00	$3.0000	/line	$3.00	$3.0000	$0.00	0%	$0.00
Additional 800 Fee	3.00	$9.00	$3.0000	/line	$9.00	$3.0000	$0.00	0%	$0.00
Complex Routing	0.00	$0.00	$0.0000	/line	$0.00	$0.0000	$0.00	0%	$0.00
Centrex Line Fee	0.00	$0.00	$0.0000	/line	$0.00	$0.0000	$0.00	0%	$0.00
Interstate CC Surcharge	0.00	$0.00	$0.0000	/call	$0.00	$0.0000	$0.00	0%	$0.00
IntraState CC Surcharge	0.00	$0.00	$0.0000	/call	$0.00	$0.0000	$0.00	0%	$0.00
IntraLATA CC Surcharge	0.00	$0.00	$0.0000	/call	$0.00	$0.0000	$0.00	0%	$0.00
InterState Directory Assistance	0.00	$0.00	$0.0000	/call	$0.00	$1.4000	$0.00	0%	$0.00
IntraState Directory Assistance	0.00	$0.00	$0.0000	/call	$0.00	$0.5000	$0.00	0%	$0.00
International Directory Assistance	0.00	$0.00	$0.0000	/call	$0.00	$0.0000	$0.00	0%	$0.00
Directory Assistance Completed Call	0.00	$0.00	$0.0000	/call	$0.00	$0.0000	$0.00	0%	$0.00
Tax	0.00	$0.00	$0.0000	/$	$0.00	$0.0000	$0.00		$0.00
Discount		$0.00	0.0000	%			$0.00		
Total	120,510.41	$6,288.46	$0.0522	/min	$4,098.87	$0.0040	$2,189.59	35%	$26,275.03

Exhibit 1.   Automated Variance Analysis: Billed Rates versus Contracted Rates
(Courtesy of Teligistics.)

for a monthly fee, Teligistics will take the client's bill from the carrier, run it through an automated error detection system, and then send it to the client for payment (or the bill may be paid on behalf of the client). A sample variance report provided to the client is shown in Exhibit 1.

Special Needs and Groups within the Organization

Like Orwell's pigs, some groups are clearly more equal than others in terms of their telecom needs. When developing a comprehensive vision of telecommunications — how costs are to be minimized while maintaining service levels — all special groups need to be considered. The classic example is the call center. With hundreds or even thousands of agents, call centers (also called contact centers) are massive bandwidth and service users. Uptime is essential and in some cases, such as Dell Computer Corporation, telecommunications provides the sole "face" of the company to the end customer. If the phone lines and Internet cables are down, how can computers be ordered?

Tailoring of requirements helps in negotiations and architectural design. If electricity traders make 50 percent of their profits in just 5 percent of the available trading day, the phone lines really need to be up 99.999 percent of the time. Hence, additional circuits, rerouting features, and other contingency services need to be included in any negotiations with the local or long-distance carrier. If the contract negotiator fails to appreciate specialty group requirements, long-term telecom costs could be inadvertently increased.

Some other considerations that affect the cost management vision include:

- *Is telecom decision making centralized or decentralized?* The preference for centralized-decentralized operations swings along its arc every decade or so. However, for telecommunications cost management, centralization has always been best. Carriers reward volumes and a balkanized approach to telecom always means higher cost. "Boudreau" in Beau Bridge, Louisiana, may get a good Frame Relay price from his brother-in-law; it may, in fact, be better than the corporate office in Houston was able negotiate with the carrier of choice. But considering the sum of all telecom costs, the corporate agreement is most likely less expensive.
- *Is change constant?* If so, long-term contracts are even more risky.
- *Do the voice and data people talk with each other?* IT, data communications, and voice communications should be integrated. Otherwise, sub-optimization will result.

Incidental Revenue

The best way to reduce telecom costs is to find ways to make them go below zero — in other words, collect revenue from telecom-related functions. For example, one Midwestern department store chain operates a 900 number service that charges firms that call to verify prior employee work history. Telephone services for students have long been a revenue source for universities. Businesses have become increasingly clever in using telecommunications for profit, or at least offloading some of the costs to customers.

Preliminary Information Gathering

After developing a vision or goal for telecom expense management, the next step is to gather organization-specific information. With this order-of-magnitude data, decisions can be made about project development, outsourcing, auditing, and other possible actions.

The first, although not easiest, step is to develop an approximate annual spend for telecommunications in major categories. If only 1 percent of the spend is in long-distance charges and 35 percent is in cellular, the conclusion is obvious: use project time to reduce cellular expenses.

Exhibit 2 shows the expenditures of a small, California-based energy firm.

Note that the expenditures for this firm are unusual because local expenses are larger than long-distance expenses. In addition, cellular usage is high, suggesting that a judicious renegotiation/rationalization would likely pay big dividends.

Before plunging into the numbers, it is useful to gather as much strategic material as possible. By recording the following information, even if it is done at a cursory level, subsequent cost management projects will be more focused and efficient.

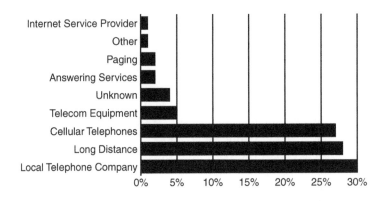

Exhibit 2. Annual Telecom Spend for a Small Energy Firm

- Organization
  - Contact information: key employees, contractors and suppliers; e-mail IDs, Web sites, etc.
  - Structure (voice, data, centralized, decentralized, etc.), organization chart
  - Functional responsibility for project management. How is it structured?
  - Staffing
  - Security awareness (who is responsible; organizational perspective)
- Strategy
  - Technology strategy
  - Customer strategy versus employee/internal requirements
  - Growth/trends
  - Buildout plans
  - Geographic issues
  - Business alignment (network and infrastructure with business needs)
  - Outsourcing plans or opportunities
  - Specific (already developed) business plans and requirements
- CRM (Customer Relationship Management)
  - Characterize the customer (profile)
  - What architecture supports CRM now (call centers, hardware, software, network)
  - Volumes of usage (now versus historical)
  - The general process (customer calls or e-mails; asks for x, then y happens, etc.). How disputes/complaints are resolved (e.g., trouble management system)
  - Describe elasticity of demand (will customer buy more product or services if service is outstanding?)
  - What are the trends for CRM (volumes, changes in entry point such as Web versus phone, changing nature of customers, etc.)
- Process
  - Ordering and provisioning (external and internal includes moves, adds, changes)
  - Bill payment
  - Maintenance of equipment and software
  - Project management for large communications projects
  - Policies and procedures

- – Escalation process
- – Asset management
- – Vendor management
- ■ Risk management
- ■ Cost management
  - – Budget
  - – Contracts and terms (carriers, equipment vendors, telecom software vendors)
  - – General ledger (G/L) and accounts payable (A/P) information
  - – Capital and expense management
  - – Circuit/equipment inventory
  - – Carrier reports
  - – Analysis of billing (tariff compliance, exception reports)
  - – Traffic/call accounting
  - – Chargeback
  - – Performance and optimization
  - – Network management tools
  - – Traffic analysis/measurement tools
  - – PBX/VM configurations
  - – Network topology and diagrams
  - – Voice and data services, including IVR/CTI

The topics are relatively high level and may or may not be directly pertinent or quantifiable as to cost-saving opportunities. However, many qualitative improvements have financially positive results, which can be measured later. Organizations with existing documentation on the above topics are often at a significant advantage and can move toward greater efficiency more quickly than many others.

Getting the Numbers — First Cut

In later chapters, the full methodology for telecom billing audits and wide area network reviews will be developed. Initially, however, a rough estimate of spend and a look at some telecom reports will provide enough quantitative information to start the expense minimization process.

To start, determine what time frame to analyze; a three-month period is usually more accurate than a single month because telecom expenses incurred will not necessarily be on that same month's invoice. Many providers bill fixed amounts in advance, current usage for the period, and then post usage discounts applied on the next bill through itemized adjustments. Using spread-sheets or a simple database, organize the data by telecom provider.

Next, look at the financial accounting system reports from the general ledger and subsidiary systems such as accounts payable. Unfortunately, most firms' chart of accounts and sub-ledgers do not capture telecom expenses in a way that allows them to be easily and automatically extracted. For example, employee cellular and paging costs may be individually expensed rather than via an enterprisewide invoice from a national supplier.

To compensate for the lack of "tags" in the general ledger or accounts payable files, a summary by vendor allows a manual review of expenses and categorization by type of telecom expense. Exhibit 3 illustrates a summary of payments by telecom-related vendor (the data is shown sorted by both vendor name and service provided). From the above information, a graph like that shown in Exhibit 2 can be constructed to show the largest expenses. This graph highlights the fruitful areas of review. If a firm has a way to generate revenue from telecom, it would be useful to show it here as a negative number.

Exhibit 3. Summary of Payments to Telecom-Related Vendors, per Accounts Payable System

Sorted by Vendor Name			Sorted by Service and Amount		
Vendor	Amount	Service	Vendor	Amount	Service
A-1 Answering Service	$1,468	A	Signius	$5,563	?
Accurate Messages	$4,141	A	Strategic Products	$4,572	?
Airtouch Cellular	$10,468	C	Citizens Communications	$3,542	?
Airtouch Paging	$3,556	P	Caldabaugh Comm.	$2,544	?
Alltel	$17,245	LD	Quantumlink Comm	$2,444	?
Alltel Mobile	$2,728	C	Bainbridge Comm.	$2,206	?
American Telephone Tech.	$2,009	?	Your Telephone Man	$2,104	?
Ameritech	$2,377	LEC	American Telephone Tech.	$2,009	?
Answer America	$1,505	A	Sierra Telephone	$2,001	?
Answer Xact	$3,274	A	Network Plus	$1,793	?
ARCH	$14,424	P	Century Tel	$1,649	?
AT&T	$184,736	LD	Peter J. Haller	$1,607	?
AT&T Wireless Services	$64,254	C	Procall	$1,602	?
Bainbridge Comm.	$2,206	?	Central Vermont Comm	$1,586	?
Bell Atlantic	$95,350	LEC	Conectiv Comm.	$1,459	?
Bell Atlantic Mobile	$29,406	C	Metrocall	$4,342	A
Bell South	$83,616	LEC	Accurate Messages	$4,141	A
Bell South Mobility	$4,477	C	Answer Xact	$3,274	A
Business Cell Systems	$1,530	C	Byrnes Message Bureau	$2,840	A
Byrnes Message Bureau	$2,840	A	Red Bank Answering	$1,520	A
Caldabaugh Comm.	$2,544	?	Answer America	$1,505	A
Carolina West Cellular	$1,920	C	A-1 Answering Service	$1,468	A
Cellular One	$10,590	C	Nextel	$105,571	C
Central Vermont Comm	$1,586	?	AT&T Wireless Services	$64,254	C
Century Tel	$1,649	?	Bell Atlantic Mobile	$29,406	C
Cincinnati Bell	$4,655	LEC	GTE Wireless	$12,056	C
Citizens Communications	$3,542	?	Cellular One	$10,590	C
Commnet Cellular	$2,854	C	Airtouch Cellular	$10,468	C
Conectiv Comm.	$1,459	?	US Cellular	$8,563	C
Frontier Cellular	$3,841	C	Bell South Mobility	$4,477	C
Frontier Telephone	$1,815	LEC	Frontier Cellular	$3,841	C
GTE	$44,784	LD	Commnet Cellular	$2,854	C
GTE Wireless	$12,056	C	Alltel Mobile	$2,728	C

Exhibit 3. Summary of Payments to Telecom-Related Vendors, per Accounts Payable System (Continued)

	Sorted by Vendor Name			Sorted by Service and Amount		
Vendor	Amount	Service	Vendor	Amount	Service	
Lease Corp America	$2,177	O	Southwestern Bell Wireless	$2,701	C	
Lucent	$46,040	E	Carolina West Cellular	$1,920	C	
Magnum Electronics	$1,469	E	Map Mobile Comm	$1,702	C	
Map Mobile Comm	$1,702	C	Business Cell Systems	$1,530	C	
Metrocall	$4,342	A	Lucent	$46,040	E	
Network Plus	$1,793	?	Magnum Electronics	$1,469	E	
Nevada Bell	$3,805	LEC	UUNET	$6,637	I	
Newcourt Leasing	$5,190	O	AT&T	$184,736	LD	
Nextel	$105,571	C	GTE	$44,784	LD	
Pacific Bell	$34,672	LEC	Sprint	$22,792	LD	
Peter J. Haller	$1,607	?	Alltel	$17,245	LD	
Procall	$1,602	?	Bell Atlantic	$95,350	LEC	
Quantumlink Comm	$2,444	?	Bell South	$83,616	LEC	
Red Bank Answering	$1,520	A	US West	$42,545	LEC	
Sierra Telephone	$2,001	?	Pacific Bell	$34,672	LEC	
Signius	$5,563	?	SNET	$10,208	LEC	
SNET	$10,208	LEC	Cincinnati Bell	$4,655	LEC	
South Central Bell	$1,499	LEC	Nevada Bell	$3,805	LEC	
Southwestern Bell	$1,869	LEC	Ameritech	$2,377	LEC	
Southwestern Bell Wireless	$2,701	C	Southwestern Bell	$1,869	LEC	
Sprint	$22,792	LD	Frontier Telephone	$1,815	LEC	
Strategic Products	$4,572	?	South Central Bell	$1,499	LEC	
Taconic Telephone	$1,496	LEC	Taconic Telephone	$1,496	LEC	
US Cellular	$8,563	C	Newcourt Leasing	$5,190	O	
US West	$42,545	LEC	Lease Corp America	$2,177	O	
UUNET	$6,637	I	ARCH	$14,424	P	
Your Telephone Man	$2,104	?	Airtouch Paging	$3,556	P	
Total	$951,389			$951,389		

$951,389 (77.8%) out of $1,223,107 for Quarter 9/26/99 to 12/25/99

Summary and Legend			%	Estimated Amounts
A	= Answering service	$19,090	2	$108,000
C	= Cellular provider	$262,661	28	$1,512,000
E	= Equipment provider	$47,509	5	$270,000
I	= ISP	$6,637	1	$54,000
LD	= Long distance	$269,557	28	$1,512,000
LEC	= Local access	$283,907	30	$1,620,000
P	= Paging service	$17,980	2	$108,000
O	= Other	$7,367	1	$54,000
?	= Unknown	$36,681	4	$216,000
Total		$951,389		
			Estimated Annual Costs	$5,400,000

For the initial look at expenses, some of the following reports and forms can be useful. Just having a few facts in hand can be most helpful when deciding on how best to set up the full-scope cost management project (described in later chapters).

Sample Reports and Forms

There are many ways to look at available information to uncover potential cost savings. The key is to create work forms that allow important information to be easily and logically recorded so savings potential can be easily identified. Some information sources include:

- Call accounting system. Using information directly from the telephone system, call accounting reports provide details on who called where and for how long and at what cost. Small telephone systems (PBXs or smaller "key" systems) may not have this system.
- Carrier-supplied CDs with call detail
- Paper invoices (tedious to review but the detail is there in some form)
- Other sources, such as CSRs (customer service records)
- Surveys of employees. Surveys can sometimes reveal patterns and needs that are not readily perceived at headquarters. For example, users at field locations may be paying needlessly high rates for cellular telephone usage because they are not aware of better rates from the corporate plan. Also, inventory information may not be available from any other source.

Exhibits 4 through 7 show sample reports and templates, appropriate for a first-cut analysis of expenses.

The report in Exhibit 4 quickly identifies problem areas. Phone calls that last longer than a few hours often indicate technical problems, hacker infiltration of the PBX, or really long conference calls. For example, a few years ago at Enron Corp. in Houston, Texas, the telecom director was reviewing a long-duration phone call report and noted some odd data. Six calls, all to the same location (from Houston, Texas, to London, United Kingdom), all over 24 hours, and all exactly the same duration, appeared on the report. Clearly, something was amiss. After investigation, he found that a videoconferencing session, which bonded six telephone lines to achieve more bandwidth, had been inadvertently left on after the conference was over. Controls were subsequently put in place to ensure the sessions were shut down at the proper time.

Exhibit 7 documents findings by category (some are control only; most are directly expense related). By ranking the findings, attention can be devoted to the most important findings first.

Typically, these preliminary reports and templates will be rolled into a full-scope telecom cost management plan. The key point is to have some idea of potential savings and current state before starting a large project or negotiating with third parties.

Exhibit 4. Long-Duration Calls, AT&T Billing Edge

CALL_DATE	Originating State	Terminating City	Terminating State	Hours Connected	Gross Amount
07-Oct-99	NJ	TALLAHASSEE	FL	9.8	$53.81
01-Oct-99	NJ	ORANGE	NJ	5.4	$44.77
20-Oct-99	KY	PARIS	TN	5.0	$32.49
21-Oct-99	KY	PARIS	TN	4.6	$29.86
25-Oct-99	FL	LAKE CITY	FL	3.8	$38.05
28-Oct-99	FL	LAKE CITY	FL	3.5	$35.16
06-Oct-99	FL	LAKE CITY	FL	3.5	$34.91
01-Oct-99	FL	LAKE CITY	FL	3.3	$33.15
11-Oct-99	NC	BOONE	NC	3.1	$35.19
22-Oct-99	FL	LAKE CITY	FL	3.0	$29.70
18-Oct-99	NC	BOONE	NC	2.9	$32.71
20-Oct-99	NC	BOONE	NC	2.7	$30.81
18-Oct-99	MO	DIAL CONF	ZZ	2.7	$112.00
11-Oct-99	FL	LAKE CITY	FL	2.6	$25.61
13-Oct-99	NC	BOONE	NC	2.5	$28.75
19-Oct-99	NC	BOONE	NC	2.4	$27.18

BTN	Call Date	Dialed Num	Fr Place	Orig Num	Term Num	To Place	To Place	Total minutes	Gross Amount
	07-Oct-99		NJ			TALLAHASSEE	FL	587.3	$53.81
	01-Oct-99		NJ			ORANGE	NJ	324.4	$44.77
	20-Oct-99		KY			PARIS	TN	300.8	$32.49
	21-Oct-99		KY			PARIS	TN	276.5	$29.86
	25-Oct-99		FL			LAKE CITY	FL	229.2	$38.05
	28-Oct-99		FL			LAKE CITY	FL	211.8	$35.16
	06-Oct-99		FL			LAKE CITY	FL	210.3	$34.91
	01-Oct-99		FL			LAKE CITY	FL	199.7	$33.15
	11-Oct-99		NC			BOONE	NC	187.2	$35.19
	22-Oct-99		FL			LAKE CITY	FL	178.9	$29.70
	18-Oct-99		NC			BOONE	NC	174	$32.71
	20-Oct-99		NC			BOONE	NC	163.9	$30.81
	18-Oct-99		MO			DIAL CONF	ZZ	160	$112.00
	11-Oct-99		FL			LAKE CITY	FL	154.3	$25.61
	13-Oct-99		NC			BOONE	NC	152.9	$28.75
	19-Oct-99		NC			BOONE	NC	144.6	$27.18
	12-Oct-99		NC			BOONE	NC	139.5	$26.23
	07-Oct-99		NC			BOONE	NC	138.4	$26.02
	05-Oct-99		NC			BOONE	NC	134	$25.19
									701.59

Note: Criteria: duration >120 minutes. These are unusually long individual calls that may indicate either a technical problem (e.g., "stuck modem") or a billing error. The source of the problem should be investigated to prevent future occurrences.

Exhibit 5. Survey Results Form: Cell Phones, Pagers, and Two-Way Radios

Manager and Location	Number of Cell Phones	Number on Corp. Agreement	Number of Pagers	Number on Corp Agreement	Number of Two-Way Radios
Jones/Memphis	6	0	6	0	4
McNew/Phoenix	4	4	18	0	0
Lakin/LaFollet	10	10	10	0	0
Beverly/Chicago	5	0	6	6	0
Rodriguez/San Antonio	1	1	0	0	4
Proulx/ Newfoundland	2	2	0	0	0
Faulkner/Oxford	8	0	12	0	0

Summary

As the first President Bush noted, the "vision thing" is essential. By developing a vision of the outcome and gathering some preliminary facts about the telecom environment, expense management can be planned and implemented efficiently. A dividend that accrues from some preliminary fact gathering is general enthusiasm for the project. In some cases, merely a reasonably accurate graph of telecom expenses will serve as the catalyst for change.

Exhibit 6. Survey Results Form: Cell Costs, PCs, and Phone Usage

Manager and Location	Cellular Rate	How Many PCs and Servers?	How Many Hours a Day Are Used on PCs?	How Many Hours a Day Are Used on Telephones?	How Many Phone Lines Are Used? How Many Telephones Use Those Lines?	What Is the Percentage of Abandoned Calls?
Site 1	$210 per month for 6 phones	2 PCs and 1 server	8	8	1 phone line and 2 telephones	2 or 3
Site 2	N/A	6 PCs and 1 server	6	4	5 phone lines and 10 telephones	5
Site 3	N/A	3 PCs and 1 server	8	8	3 phone lines and 6 telephones	None
Site 4	$300.00 per month for 5 phones	3 PCs and 1 server	6	6	3 phone lines and 5 telephones	None
Site 5	$50.00 per month for 1 phone	3 PCs and 1 server	6	6	4 phone lines and 4 telephones	10
Site 6	$50.00 per month per phone	7 PCs and 1 server	8	8	5 phone lines and 9 telephones	15
Site 7	N/A	3 PCs and 1 server	6	6	9 phone lines and 3 telephones	1 or 2
Site 8	$50.00 per month per phone	5 PCs and 1 server	8	2	4 phone lines and 3 telephones	10

Exhibit 7. Template for Recording Detailed Findings

Assessment Category	Topic	Current Status	Risk (H, M, L)	Resource Requirement (H, M, L)	Recommendation for Improvement
Organization	Staffing	Only one employee is responsible for all LD voice services and technology; same issue for pager and cellular administration	H	M	The exposure to firm is great as no backup support or continuity exists in these support areas. In addition, a skills gap analysis should be conducted. Develop detailed procedures for the PBX/voicemail, pager, cellular functions.
Strategy	Technology	Planning to redesign the network architecture, converge voice and data over the same circuits, move to VPN and then Frame Relay technologies; looking at E-business opportunities, possible Web hosting, data warehousing, and move toward a vendor-managed inventory; no formal implementation plan exists	M	H	Rather than planning for building a future network architecture first and then looking at today, recommendation would be to reverse the two, which should reveal alternative solutions for the technology path the firm is taking.
Customer relationship management	Customer profile	The customer satisfaction group has segmented the customers into five markets/elasticity of demand (e.g., convenience user, price shoppers); performs "mystery shopping," conducts interviews, etc.	H	L	Seems to be a communication disconnect between different markets and among firm's employees. CRM findings on this and other topics should be disseminated more frequently.

Process	Escalation	Plan is to implement "Solution 1" software for trouble resolution as well as asset management but we did not see a formal document and project timeline	H	M	Consider using Solution 1 capability to be telephony enabled.
	Security	Not clear that when an employee is terminated, a process exists to notify administrators	H	L	For NT, mainframe, and PeopleSoft, ensure proactive HR process to notify the administrators.
	Controls and policy	No retention policy on voicemails and e-mails; firm is potentially exposed if there is ever a "legal discovery" situation	H	M	Awareness and corporate policy enforcement process.
Cost management	Expense management	There are no standards in place for telecom products and services. For example, only a few people are on the one-rate cellular plan. Another example is that although pager prices are good ($4), there are no controls on pager options. In fact, all regions do not use the national account for their pagers. Also, many are on the national account and others are not, as employees maintain that local services are cheaper and expense them so there is no handle on cost. Many employees expend more than $80 a month on cellular. There are 400 cellular phones being billed by ABC Company at a monthly cost range of $26K to $31K. Last month usage was 82,567 minutes for an average of 206 minutes per phone. We are not aware of the other service providers being used at this point.	H	M	Evaluate required services by store/region/etc. Also, who can and/or should have both phone and pager and other telecom services and equipment. If standardized, firm can leverage usage on carrier and supplier contracts. For example, over $80 a month (possibly over $40 category as well) should be converted to a cheaper fixed rate plan. An incentive for store buy-in may be an approach like 50 percent of total savings will be passed back to the stores.

Exhibit 7. Template for Recording Detailed Findings (Continued)

Assessment Category	Topic	Current Status	Risk (H, M, L)	Resource Requirement (H, M, L)	Recommendation for Improvement
Cost management	Bill audit and review process	The auditing and processing of telecom invoices is entirely manual. For example, for pagers and cell phones, if one of the stores requests, there is a greater than $200 report that can be printed. Does not appear that much checking (e.g., sampling) is done on billing invoice accuracy.	H	H	Firm should evaluate using computerized workflow and analysis tools to automate the invoice processing function. In any event, potential savings exist in conducting a billing and recovery audit today in areas such as rating accuracy; service charges and waivers, calling plan evaluation, contract evaluation, etc.
	Contracts	The current X and Y agreements have a total revenue minimum requirement as well as sub-minimums for key services such as VNET, Frame Relay, audio conferencing, toll-free services, and international traffic. At one point, the sub-minimum for audio conferencing was not met and the contract had to be amended.	H	M	Firm should consider structuring an agreement that focuses on total revenue (or total volume commitments) rather than many sub-minimums
		Rates are not tiered to provide increasing discounts as volumes of minutes, ports, circuits, etc, increase	M	H	Firm should perform an analysis to determine if a tiered rate structure can provide deeper discounts as volumes increase

Category	Sub-category	Finding	Risk	Risk	Recommendation
Performance and optimization	Voice traffic	Voice trunk (T1) studies are not conducted periodically to ensure they are neither over-utilized (busy signals) or under-utilized (excess cost)	M	H	Implement an automated or regular process to obtain and evaluate utilization
		No reporting mechanism in place to break down traffic (e.g., IntraLATA, interstate, calling card) and to provide trending data	M	H	Develop tools/processes to capture and analyze data
Security and business continuity	PBX	PBX security configuration and penetration reviews are not conducted on a regular basis	H	M	Firm should periodically review the controls over the voice network in order to validate prescribed configurations and their overall effectiveness
Service levels	Network	Service levels may not be adequate for the requirements of the business (data)	M	M	Given the increasing reliance of digital circuits and monitoring platforms, firm should consider requiring service levels in excess of those in the current contract

Risk Ratings   *Definition*

High   Finding, if not addressed, could result in significant loss (opportunity or actual loss) or exposure to operations disruption.

Medium   Finding, if not addressed, could result in moderate level of loss or exposure to operations disruption.

Low   Finding will not significantly affect costs or operations, but should be addressed as part of a comprehensive response.

Resources   *Definition*

High   Technical or subject-matter knowledge extensive and requires significant time to implement.

Medium   Average technical or subject matter knowledge and requires moderate time to implement.

Low   Minimal technical or subject matter knowledge and minimum duration project.

Chapter 3

Telecommunications Auditing

The man who makes no mistakes does not usually make anything.

— Edward John Phelps, 1889

Employees involved with paying telecom invoices can attest to the complexity of reviewing bills, the arcane terminology of telecom services, and the plethora of billing options available to business customers. It is no wonder that with limited resources and the lack of skilled bill auditors, many businesses simply direct their Accounts Payable (A/P) department to pay the invoices without adequately reviewing them. Unfortunately, such practices lead to significantly higher business operating costs. In many cases, telecom expenses are the fourth largest expense after payroll, cost of goods sold, and facilities.

External bill auditors have traditionally identified a 5- to 30-percent savings on telecom bills by simply identifying rating errors, double billings, bad/uneconomic contracts, unused circuits, unauthorized services, and other errors. These issues are primarily caused by inadequate controls and decentralized management over telecom processes, as described later in this chapter.

The following sections describe the basic process for auditing telecom expenses. First, we discuss the reasons for telecom billing discrepancies. We then identify the nine common steps involved in ordering, installing, and paying for telecom services. Finally, we provide details of how reviewing bills will reduce telecom costs. Appendix H includes a telecommunications audit program designed for the individuals performing the detail work.

Reasons for Telecom Bill Discrepancies

Telecom service providers, their competitors, and business customers all play a role in inaccurate bills. Let us start with the pivotal 1996 Telecommunications Act.

Telecom Service Providers

The Telecommunications Act of 1996 dramatically altered the telecom landscape by deregulating telecommunications. The resulting fierce price competition in long-distance rates, which drove down telecom service providers' revenues, prompted a decline in customer service levels that often leads to inaccurate billing.

Historically, a carrier's customer account manager may have supported a few major business accounts. However, with the pressure to control costs and boost profitability, the same account manager now services more accounts with less time for each one. Turnover and the lack of adequate training may also contribute to declining service levels.

For example, one long-distance provider's sales representative insisted to a client that his firm did not offer toll fraud insurance. He was somewhat embarrassed when shown the details of a toll fraud offering on his firm's public Web site. Telecom account managers must understand their customer's business and telecom usage, as well as their own offerings. Otherwise, service orders may not be correctly executed and billed.

Overly Aggressive Competitors

The Telecommunications Act of 1996 ushered in new competitors eager to aggressively increase their market share in the local and long-distance markets. The terms "slamming" and "cramming" were quickly added to the telecom lexicon as some service providers allegedly engaged in illegal business practices to gain new customers.

Slamming

Slamming is the illegal practice of switching a company's preferred local or long-distance service provider without explicit authorization. When a company orders telephone lines from the local telephone company, it specifies the preferred long-distance carrier for each line. The preselected long-distance carrier for a telephone line is commonly referred to as a PIC (preferred interexchange carrier). In telecom vernacular, we say that the line has been "PICed" to a particular carrier.

Telephone customers are slammed in a variety of ways. A common method is the use of forged copies of Letters of Authorization to the local telephone company, which "authorize" switching the PIC to an unauthorized provider. In another method, a service provider contacts a customer about new services but does not inform the customer that selecting the new service will also result in changing the preferred long-distance provider or PIC. In some cases, particularly for residential services, truly deceptive practices have been used. For example, "free" raffle tickets at retail malls have tiny print at the bottom that authorizes a switch from one carrier to another. When unsuspecting victims fill out and sign the raffle ticket, they are unknowingly authorizing a carrier change.

Cramming

Cramming is the illegal practice of adding charges to a business telephone account for products and services that have not been authorized. In one press release by the Federal Communications Commission (FCC), a service provider was fined for placing "unauthorized fees for 'membership' in the 'Friends to Friends' psychic services hotline and 'other' charges on consumers' telephone bills." What the FCC found particularly egregious about these violations was that many customers were billed for these services although they had no contact with the service provider or the psychic services hotline.

PricewaterhouseCoopers has encountered several cases of cramming in its bill audits. While auditing bills for a global professional services firm, there was one office with a telephone line that was billed twice for voicemail — by two different service providers. If a representative from either service provider had called the telephone number prior to cramming the line, he would have found that the line already had voicemail — from an onsite Avaya voicemail system that the firm owned. Another common example of cramming is a charge for "inside wiring," which is, in most cases, an unnecessary line maintenance fee.

FCC Enforcement Actions

Antislamming/cramming enforcement actions have been taken by several Public Utility Commissions (PUCs) — a state's regulatory body — and the FCC against service providers that allegedly engaged in these illegal activities. As Exhibit 1 illustrates, between February 2000 and December 2001, the FCC imposed fines, entered into Consent Decrees or issued Notices of Apparent Liability that totaled approximately $15,820,000. Note that some of these service providers voluntarily brought the slamming issues up to the FCC and began proactive remedial measures.

Business Customers

One major business issue that corporations face today is how to react to business cycles and rapidly changing economic conditions. Corporations may engage in mergers, acquisitions, and right-sizing activities. Without adequate telecom cost controls, businesses may drive up their total telecom costs, which hurts the IT budget and the earnings before interest, depreciation, taxes, and amortization (EBIDTA).

Exhibit 1.   FCC Antislamming Enforcement Actions

Date	Company	Action	Amount
12/17/01	America's Tele-Network Corp.	Fine	$1,020,000
09/10/01	All American Telephone, Inc.	Fine	$920,000
04/17/01	AT&T Communications, Inc.	Fine	$520,000
12/07/00	Business Discount Plan, Inc.	Fine	$1,800,000
12/07/00	Coleman Enterprises, Inc.	Fine	$750,000
10/23/00	Vista Services Corporation	Fine	$680,000
07/21/00	Qwest Communications International, Inc.	Consent Decree	$1,500,000
06/06/00	MCI WorldCom	Consent Decree	$3,500,000
04/25/00	Excel Telecommunications, Inc.	Consent Decree	$400,000
03/27/00	Sprint Communications Company, LP	Consent Decree	$250,000
03/02/00	Brittan Communications International Corp.	Fine	$1,120,000
02/17/00	Long Distance Direct Holdings, Inc.	Notice of Apparent Liability	$2,000,000
02/09/00	Amer-I-Net Services Corporation	Notice of Apparent Liability	$1,360,000

Courtesy of the Federal Communications Commission, www.fcc.gov, http://www.fcc.gov/eb/tcd/slam.html.

Reacting to Cyclical Business Activities

Business expansions and contractions involving significant changes to employee headcount directly impact telecommunications costs. Typically, these business activities result in overpayment for unused circuits and inappropriate services.

In an expansionary period, a company providing services to its new employees may incur significant expenditures for installing lines to the employee's desk, purchasing hardware such as additional cards for the PBX, or provisioning additional trunks from the telephone company. Services such as call forwarding may be inappropriately provided to employees who staff inbound contact centers. Employee telephone abuse is frequently attributed to call-forwarding features that allow employees to forward toll-free phone calls from friends and family to the employee's home after business hours.

In an optimal control environment, appropriate controls are implemented to ensure new telecom assets, and services and telephone features are authorized and commensurate with job responsibilities. Replacing antiquated, manual chargeback systems with automated, scalable systems provides an additional level of control. Employees and their cost center managers can monitor their own network, calling card, and long-distance usage, and report fraudulent activity to appropriate personnel.

During economic contraction, when the organization typically reduces headcount, telecom assets such as cell phones, pagers, calling cards, and radios may not be recovered. Also, services may not be disconnected appropriately.

Experience shows that ineffective asset management contributes to losses. The exit interviewer may not have objective information on the departing employee's telecom assets (cell phone, pager, calling card, etc.); sometimes, information from Human Resources is not current. The net result is that services may continue to be provided to the terminated employee for months after termination.

Effective controls ensure that any reduction in workforce will trigger a set of actions to identify and recover assets and remove services. Without adequate controls, cost centers could be inaccurately billed for usage and equipment charges; significant business risks are incurred from disgruntled individuals misusing or compromising telecommunications services and systems.

Consolidating Offices after Mergers or Acquisitions

Companies that merge with or acquire another entity typically relocate or consolidate offices. Without appropriate bill review processes, consolidation activities frequently lead to paying for unused services and dangling circuits — circuits that are not terminated at one endpoint — because they have not been removed from the telephone company's billing records.

Although the local telephone company has disconnected the enterprise's circuits, the long-distance carrier can still render usage charges. The enterprise is essentially paying for someone else's long-distance services. This situation occurs when the local telephone company reassigns the circuit to another enterprise. If the circuit has not been removed from the original enterprise's long-distance carrier's database, the long-distance company will continue to bill the original enterprise for all usage charges incurred by the new circuit owner.

Implementing Technology Solutions

The advent of the Internet, intranets, and extranets has placed increased demands on network bandwidth and availability. Enterprises are upgrading voice and data infrastructure to enable Customer Relationship Management (CRM) solutions, E-business, and other strategic initiatives.

Customers expect a prompt response, whether they are purchasing by telephone or the Internet. If Web-enabled transactions slow to a crawl, customers will buy from a competitor's site. CRM technology investments are unsuccessful if the most profitable customers get busy signals from the contact center or encounter an auto-attendant nightmare. The enterprise will most likely lose the sale — and possibly the customer.

Companies competing for mind share with today's sophisticated consumer must continue to improve the quality of the customer's experience with the contact center. Today, the customer's attention span is shorter than ever. Customers have more choices, easier access to information, and higher expectations of service and availability.

In response to these concerns, companies traditionally increase bandwidth without appropriate consideration of costs. That is, they may hurriedly throw

Exhibit 2.   Nine Steps in the Telecom Service Ordering Life Cycle

Step	Process	Description
1	Request service	An end user requests a cell phone, pager, telephone extension, or other type of telephone service.
2	Authorize service	The end user's manager approves the service. In situations where chargeback systems are deployed, the manager identifies the appropriate cost center to allocate the one-time equipment or installation charge and the recurring service charge.
3	Order service	The request is sent to the Telecom department, where a telecom analyst orders the service from the service provider.
4	Provision service	The service is added by the service provider.
5	Validate service	The Telecom department confirms that the service was installed.
6	Invoice receipt	A/P receives the invoice from the service provider and inputs the information in the financial system.
7	Invoice authorize	A/P sends the invoice to the Telecom department to authorize payment.
8	Invoice analysis	The Telecom department reviews the invoice and approves payment. The invoice is rerouted to A/P.
9	Invoice payment	A/P pays the invoice.

excess bandwidth at the problem rather than taking the time to adjust in proportion to actual need. The need for more capacity, more services, and more fault tolerance capabilities must be balanced with the need to control costs. Too much capacity leads to excessive costs. In a recent audit, one enterprise added more than a dozen long-distance T1s as a contingency for Y2K; six months after the millennium change, the excess T1s were still in place.

Adding services without appropriate capacity planning can result in paying for unused circuits and services. Asset management systems that inventory line, circuit, and hardware assets, coupled with real-time monitoring of network and trunk utilization call accounting reports, will help control over- and under-trunking.

The Telecom Service Ordering Life Cycle

To better understand why organizations are increasingly challenged to account for telecom costs, one needs to look at the telecom service ordering life cycle. Exhibit 2 depicts a typical company and the nine different steps involved — from requesting service to paying the telephone bill. In this example, the service request process starts with a departmental end user.

In reality, the process is much more complex than outlined above. As Exhibit 3 shows, many functional groups could request, order, or review bills. For example, the origination point for requesting telephone services could begin with the

Exhibit 3. Functional Groups Involved in the Telecom Ordering Life Cycle

Process	Telecom Dept.	Contractor	Purchasing	A/P Dept.	Manager	EndUser
Request service	√	√		√	√	√
Authorize service	√	√		√	√	
Order service	√	√	√	√	√	√
Provision service	√	√				
Validate service	√	√	√	√	√	
Invoice receipt	√	√	√			
Invoice authorize	√	√	√	√		
Invoice analysis	√	√		√		
Invoice payment				√		

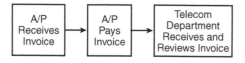

Exhibit 4. Telecom Department Reviews Invoices after Payment

Telecom Department, or in some cases, it could be outsourced to a contractor. With so many functional groups involved in the telecom service ordering life cycle, it is no surprise that companies typically overpay for services because of fundamental issues in their ordering and billing reconciliation processes.

Two Approaches to Paying Bills

A company can take one of two approaches in paying its telecom bills. In the first approach, depicted in Exhibit 4, the Accounts Payable department (A/P) pays all invoices without review and forwards a copy to the Telecom department for later review.

This approach ensures timely payment of carrier invoices and avoids service interruptions due to late payment, as most payment terms with carriers specify net 30. One disadvantage is the potential for overpayment, as the Telecom department typically runs lean on staff and may not have the resources to methodically review the invoices each period. Because the invoice is already paid, the Telecom department may not consistently review all invoices. Inertia often rules — payment is made without scrutiny if the supporting paperwork exists and the dollar amount is reasonably close to last period's payment.

In the second payment approach, depicted in Exhibit 5, A/P logs the invoices into the accounting system and sends all invoices to the Telecom department. The Telecom department reviews the invoices, notes and investigates exceptions, approves the invoices, and then forwards the invoices to A/P for payment.

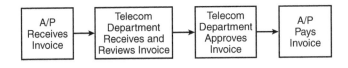

Exhibit 5.   Telecom Department Reviews Invoices before Payment

The second approach reduces the potential for overpayment, but also increases the potential of service interruptions or late penalties because the cycle time from invoice receipt to payment is doubled. The Telecom department may not have the capacity to review the invoices in a timely manner, or the invoices may get lost or misdirected in routing. Additionally, if the telecom department incorrectly disputes charges that are actually owed to the carrier, late penalties will accrue.

The second approach is generally preferred because billing errors are typically identified and resolved more quickly by the Telecom department. Companies that have their A/P departments pay the bill without the Telecom department first reviewing the bills may want to train an A/P clerk to perform the bill review. For this to be successful, the A/P clerk should have access to the Telecom department's circuit and service inventory and be aware of service connections, disconnections, or changes to validate the bills.

Components of a Telephone Bill

Every local telephone company has a unique billing format and style, but they all share similar characteristics. The first page typically provides the following information:

- Bill payment or remittance address
- The customer billing address
- The account number, including the main telephone number
- A summary of charges that includes:
  - Local monthly service (recurring monthly service charge for telephone lines and circuits billed to the account)
  - Local calls (local toll charges billed directly by the local telephone company
  - Information charges (directory assistance charges)
  - Taxes (federal, state, and local taxes)
  - Long-distance carrier billing (usage charges for long-distance services billed by a long-distance carrier on the local bill; Exhibit 6 shows AT&T billing on this local bill)

A closer look at the telephone bill in Exhibit 6 shows one problem common to all local telephone bills. Although Exhibit 6 directs the customer to Page 2 for details on the monthly service charge, the details behind the monthly service charge remain unclear (Exhibit 7). The "Monthly Service" charge of $240.84 is not clearly explained.

Exhibit 6. Example Local Telephone Company Bill

AMERITECH Bill Payment Center Saginaw, MI 49784-0003	Account Number 517 555-2445 3021
Customer 1000 Lincoln Street Anytown, MI 40000	March 22, 2001

Amount If Paid After April 12, 2001	309.17	Amount If Paid on or before April 12, 2001	306.88

Detach and mail top section with your payable to Ameritech. Write account number on check.
PLEASE ALLOW FIVE DAYS WHEN PAYING BY MAIL OR AT AUTHORIZED AGENCY.
Mail payments to: Ameritech, Bill Payment Center, Saginaw, MI 49784-0003

517 555-2445 3021

CUSTOM BUSINESS SERVICES
BILLING SUMMARY   **MAR 22, 2001**

Previous Bill	Payments Thank You	Adjustments	Balance	Current Charges	Total Amount Due
292.36	292.36	0.00	0.00	306.88	306.88

(Second Page)
SUMMARY OF CURRENT CHARGES

Ameritech Local Service
For Detailed Charges See Page 2

Monthly Service	240.84
Local Calls	18.19
Information Charges..	.00
Long Distance	7.83
Local, State and Federal Charges	14.76
Taxes (Fed 7.80)(St 15.59)	23.39
Ameritech Local Service CURRENT CHARGES..............................	**305.01**

AT&T
For Detailed charges See Page 41

Information Charges..	.75
Long Distance	.97
Taxes (Fed .05)(St .10)	.15
AT&T CURRENT CHARGES...	**1.87**
TOTAL CURRENT CHARGES	**306.88**

In fact, it is not until you turn to Page 3 that lines included in the "Monthly Service" charge can be identified. Exhibit 8 shows Page 3 of this bill. In this case, two of the telephone lines, (517) 555–2225 and (517) 555–2445, which are included in the "Monthly Service" charge, had usage charges.

If these lines did not have usage charges, they would not have appeared on the bill. The local telephone bill does not present the complete picture of what lines and services are billed to the company. It also does not tell you the service address where the lines terminate and services are delivered. The address on the front of the bill is the billing contact name and address where the service provider mails the bills. For example, if the billing contact is the

Exhibit 7.   Page 2 of a Local Telephone Company Bill

DETAILED CHARGES

Page 2

Billing Inquiries call 1-800-HELP-ME!
Service Inquiries or Orders call 1-800-HELP-YOU
Repair Problems Call 1-800-REPAIRS

IMPORTANT INFORMATION

> Our records show that you have selected AMERITECH as your presubscribed carrier for all of your IntraLATA long distance services.
> Our records show that you have selected AT&T as your presubscribed carrier for all of your InterLATA long distance services.
>
> NEW STATE CHARGE
> We want to make sure you know that a new state access charge will appear on your bill monthly starting June 11, 1999. This charge helps Ameritech recover some of the costs associated with the telephone lines used to make long distance calls. This fee per line is $0.05 for Centrex, $0.45 for single and multi-line businesses and ISDN BRI, $2.25 for ISDN PRI and ADTS/ADTS-E.

CURRENT CHARGES

Monthly Service --- Dec 22 thru Jan 21	173.40
Federal Access Charge	67.44
Total Monthly Service Charges	240.84

Exhibit 8.   Page 3 of a Local Telephone Company Bill

DETAILED CHARGES

Page 3

See Page 2 for Customer Service Telephone Numbers

Long Distance — continued

No.	Date	Time	Place Called		Number	Code	Min	
1	3-16	1101A	WEBBERVILLE	MI	517 555-4899	D	2	.50
			Total Calls Charged to 517 555-2225					3.10
			Calls Charged to 507 555-2445					
2	3-02	1158A	CHARLOTTE	MI	517 555-4333	D	3	.74
3	3-02	1218P	JACKSON	MI	517 555-5680	D	2	.53
4	3-08	225P	JACKSON	MI	517 555-6613	D	1	.27
			Total Calls Charged to 507 555-2445					1.54

Accounts Payable manager, the address on the front of the bill will be the A/P department address.

Because of these issues, companies should always request a Customer Service Record (CSR) for their local telephone accounts. A CSR describes in detail the following information:

- Billing address
- Service address

- List of individuals who are authorized to make changes to the account
- List of all telephone numbers and, under each number, a detailed list of services billed to that number

By comparing information on the CSRs to the company's line and circuit inventory, a company will be able to validate the services for which it is paying.

Sometimes long-distance carrier charges show up on a local telephone company bill in addition to the expected charges from the long-distance carrier. This occurs when the long-distance charges were not rated under the company's corporate discount plan. Sometimes, this situation occurs when a field office manager, unaware of a corporate plan, calls the local telephone company and signs a contract for services. Although the contract is with the same carrier, the rates will likely be at the "mom-and-pop" rates that are typical for smaller firms. The long-distance carrier has no easy way of identifying such "rogue" plans.

Companies contract with a long-distance carrier for long-distance services under a discount plan. The Telecom department provides working telephone numbers (WTNs) to the long-distance carrier when new lines are ordered and the local telephone company has assigned the line number. The WTNs are placed into a long-distance carrier's database, and calls are billed per the contract plan. If the phone numbers are not in the database, the correct discounts will not be applied to the call.

Sources of Billing Information

Until recently, reviewing telephone bills consisted of manually flipping through reams of paper. For a few enterprise customers with several million dollars in telecommunications spend, the bill call detail records were so gargantuan that bills were literally delivered on a palette. Forklifts would be needed to move the bills from the mailroom to the Telecom department and finally to Accounts Payable. This led many customers to request only summary bills that did not contain the detail needed to validate the bills.

Now, all carriers offer bills in a variety of formats, including magnetic tape, CD-ROM, and via the Internet. Local and long-distance carriers may charge a small fee for the electronic information, but the added flexibility in reconciling the billing detail will almost always offset the cost.

Getting Down to the Details: Audit Steps

This is the "mud wrestling" section of the book — where we get dirty with the minutia of telecom auditing. The following steps outline a basic but broad-scope audit of telecommunications expenses. Although not essential, intermediate or higher expertise in spreadsheet and desktop database usage facilitates these audit steps.

To present the material in a logical, step-by-step sequence, the process has been somewhat simplified. As mentioned, Appendix H lists the detailed steps that a field auditor needs to follow to identify and document findings at the detail level. Beyond that, each organization will have specific concerns that should be included in a detail review. Often, one of these concerns is proper allocation of costs internally, not just the total expense that goes out the door.[1]

The time required to perform a telecom review varies by size, media of records (paper versus CD-ROM), knowledge of the reviewers, geographic distribution, and general availability of organizational information. Time estimates are roughly as follows:

- Several days for initial interviews and requests for documentation; this depends on the current business environment or profile of the project
- Four-to six-week lag period to wait for carrier documentation: during this time, someone should track receipt and verify completeness as it is received
- A week or less for data entry: this can be outsourced to a temp agency to better manage project costs and enhance speed of data entry
- Two to three weeks for analysis and creation of project deliverables, depending on the types and extent of issues and opportunities

Typically, a project can be completed within a month of receipt of bills and documentation. Larger projects will require more staff on the data entry and analysis portions.

Review Contracts

Obtain copies of all telecom-related contracts, including those for voice long-distance, data communications (such as Frame Relay and dedicated circuits), local, cell phone, pager, wireless services, and others, such as VSAT links. In a spreadsheet, summarize key information that directly affects pricing. For example:

- Implementation and termination dates.
- Any ramp-up period (i.e., a point between the effective date of a contract and the point at which required minimums must be in place). Ramp-up time is often provided when a large organization is converting from one provider to another; the favorable rates are provided during the interim of the cutover.
- Per-minute rates. This is not as easy as it would appear. Interstate rates for switched-to-switched, switched-to-dedicated, and dedicated-to-dedicated may be straightforward. In contrast, intrastate rates vary considerably and may not be documented in the contract. While obtaining this information in total (for all states) may not be worth the trouble, intrastate per-minute charges for the most common calling patterns should be documented. Ideally, the contract will list a flat intrastate rate for the state(s) where most business is conducted.
- Special price schedules. Rates are sometimes tiered. For example, at a volume of up to one million minutes per month, the discount off standard tariff might be 44 percent; beyond that volume, the discount may go up to 47 percent. Watch for discounts that drop to zero.

- Minimums for all services. Some carriers make minimum commitments more complex than necessary, with a bevy of sub-minimums that require constant oversight. Document the minimum required minutes by service.
- Traffic percentages. Contracts may require that a certain percentage of the traffic be dedicated-to-dedicated or penalties will apply.
- Installation and other waivers.
- Up-front and scheduled sign-on credits.
- Any special equipment or service provisions. For example, are audio conferencing services provided at a discount?
- Any locations specifically listed for discounts, lower rates, etc. This will be useful later when examining billing by location on the invoice.

The following clauses, adapted from a carrier contract with a large firm in the Southwest, illustrate terms that might typically need to be monitored and included in the pricing summary spreadsheet.

- At least 55 percent of the customer's long-distance services usage must be rated under schedules B and/or C (meaning dedicated-to-dedicated and/ or dedicated-to-switched traffic).
- The customer must bill at least $95,000 per month in the Xtra-service Volume Pricing Plan and $23,000 per month in local channel services.
- In the first year, the customer must satisfy 66 percent of the minimum commitment with usage previously provided by another carrier.
- XYZ carrier will provide an annual credit of $78,000 to the customer's bill in the 13th, 25th, and 42nd months.
- International calling for direct-dialed services will have the following discount schedules off tariff:
  - $0 to $50,000 (19 percent)
  - Over $50,000 and up to $99,000 (21 percent)
- Usage rates are based on initial 18 seconds or fraction; additional rates apply for each additional 6 seconds or fraction.

Contracts are typically 50 to 100 pages long, so this documentation step will be time-consuming for many organizations. Carriers typically provide summary sheets to the customer when bidding for the business; they are a good starting point because they are often more straightforward than the jargon-infested legal documents.

Obtain Raw and Summarized Data

If possible, get data for up to three years previous. Billing data will be available via a Web interface (provided by the carrier), CD, or paper. If data is available only on paper, the audit will be much more time-consuming and will not identify as many potential savings.

The data should be imported into a spreadsheet or database for review. If only paper invoices are available, selected data should be keyed in or scanned (OCR) and parsed into appropriate columns. Use of a database such

Exhibit 9.   AT&T Billing Edge CD Data Imported into Excel

Account #	Location Acct #	CALLDATE	FROMNUM	TONUM
2929980994989	2322259309030	2/26/2002	203-262-9985	8229850542
2929980064622	2322202232328	2/9/2002	405-224-9929	9408554200
2929980064622	2322202232328	2/25/2002	580-889-9028	9408554200
2929980064622	2322202232328	2/26/2002	254-952-2259	9408554200
2929980064622	2322202232328	2/6/2002	925-993-5668	9408554200

FROMPLACE	FROMSTATE	FROM*CITY*ST	TOPLACE	TOSTATE	TO*CITY*ST
Chattanooga	TN	Trumbull CT	St Philip	IN	St Philip IN
Chickasha	OK	Chickasha OK	Humble	TX	Wichitafls TX
Atoka	OK	Atoka OK	Kingwood	TX	Wichitafls TX
Waco	TX	Dallas TX	Chicken Scratch	TX	Wichitafls TX
Stamford	TX	Stamford TX	Kingwood	TX	Wichitafls TX

BILLTIME	CHARGEAMT	AFTERDAMT	COST
30	0.52	0.39	0.39
30	0.34	0.30	0.30
30	0.34	0.30	0.30
30	0.34	0.30	0.30
30	0.33	0.30	0.30

Note: For display purposes, the data above is shown "wrapped" across the rows; when working with live data, all information for one call will be on one line.

as MS Access is preferable because billing records such as the detail for long-distance calling can exceed the 64,000-row limit of Excel and other spreadsheet programs. Exhibit 9 shows the format of selected columns of call detail data imported from an AT&T Billing Edge CD. There are many more fields than those shown; many are not relevant to the cost analysis.

A similar effort is required for local charges, cellular expenses, and other telecom-related charges. Following are typical documents that should be collected:

- Inventories
- General ledger information
  - Vendor names
  - Annual or monthly spending
  - Types of invoices or items coded to telephone expense codes
- Vendor invoices
  - LEC (or CLEC)
  - IXC

- Telephone equipment vendor
- Calling card
- Two-way radio
- Pagers
- Cellular
- Vendor contracts
  - Contracts with LEC or CLEC for local service
  - Contracts with long-distance providers
  - Lease agreements for equipment
  - Maintenance agreements for equipment
- Other documents from vendors
  - Customer Service Records (CSRs)
  - Physical equipment
  - Circuit inventories
- Historical information
  - Moves
  - Adds
  - Changes
  - Deletions
- Disconnects Letter of Agency (required only if outside party is performing the work; see Exhibit 10)

Once the data is in electronic format, the fun can begin — running a plethora of creative exception analyses to identify potential cost savings. The audit steps below are broken into usage categories because the analysis varies by services used.

Findings Analysis

Findings are usually grouped into long-distance and local categories. The following subsections outline some of the more important factors affecting telecom costs.

Long Distance: Voice

Using the contractual terms previously identified, any existing call accounting/management reports, and the detail data obtained from the carrier(s), perform the following exception testing:

- *Divide the cost of the call by the number of minutes to obtain a cost per minute.* Stratify the costs per minute using a histogram, such as the one shown in Exhibit 11. This data reveals that most calls are in the $0.08 range with a second peak in the rather expensive $0.25 range. The data shown is from a CD-ROM containing one month's information. Much of the data for telecommunications auditing can be analyzed using Microsoft Excel, which has numerous statistical and graphical functions built in (go-to tools, data analysis, histogram).

Exhibit 10. Sample Letter of Agency (Required if an Outside Party Does the Work)

XYZ Corp.

DATE: 2/26/02

TO: Local exchange telephone companies (SWBT)
        Interexchange carriers (AT&T, MCI)
        Other common carriers
        Equipment vendors (Lucent)

Dear Sirs:

Please be advised that we have executed a contract with PricewaterhouseCoopers, L.L.P (PwC). PwC will be performing an audit on the telecommunications services of our company. PwC will act as our agent in any and all matters related to the communications network our company has in place.

This letter authorizes PwC to act as our communications representative when dealing with your firm. We authorize PwC to obtain information or copies of all our network services.

By this letter we do hereby authorize PwC to act as our communications agent to change network carriers, to obtain telephone account information, PIC codes, telephone numbers and service locations, billing data, type of service, class of service of telephone associated with the account, and any other customer information needed.

Sincerely,

Sandra Doe
Vice President of Finance
XYZ Corp.

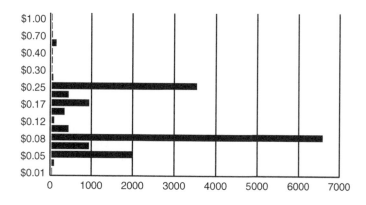

Exhibit 11.   Histogram: Number of Calls by Cost per Minute

- *Obtain a random sample of calls from a billing CD (using Excel "Rand( )" function)*. Manually calculate what the cost should be, based on the contract. Be certain to include all relevant factors, such as six-second incremental billing, setup charges for calling cards, etc. If results do not match the bill, it is important to continue the research. Telecom auditors find that, indeed, carriers sometimes do bill incorrectly, even on something as seemingly elementary as the per-minute charge.
- *Use Excel's "descriptive analysis" function to get a quick perspective on the major fields: cost per minute, maximum and minimum call durations, dates, and other fields that could affect the financial outcome.* Using the same CD-ROM data that was used for the histogram, one obtains Exhibits 12 and 13 by clicking on Tools, Data analysis, and then Descriptive analysis. The most salient information shown in Exhibit 13 is the 25-hour duration of a single call. Later analysis revealed that an employee of the organization was in the habit of leaving his laptop connected over a long-distance link in order to avoid being bothered with multiple log-ins during the day. When informed, management immediately implemented a mandatory two-hour limit on any single session.
- *Sort and summarize data by* originating *location.* This information allows the reviewer to match the volume of calls at a specific location with an understanding of the business activities conducted there. Numbers that are out of line could signify toll fraud or employee abuse, an increase in legitimate business, or a switched location that now justifies a dedicated link (or vice versa if the usage slips from previous levels). Exhibit 14 shows the results of a "from" or originating analysis of call expenditures.
- *Sort and summarize by* receiving *(to) location.* If sufficient traffic is received at a location, then it may justify a dedicated access circuit.
- *Summarize inter-office traffic.* If enough calls are made between two or more of the organization's offices or plants, dedicated circuits should be considered (traditional T1s, Frame Relay, ATM, or other links).
- *Summarize traffic by area code and country code.* Any premium numbers (900 numbers and look-alikes) will be clearly revealed, as well as possible fraudulent activity.

**Exhibit 12. Descriptive Statistics
for Per-Minute Cost**

Mean	0.118839286[a]
Standard Error	0.000675767
Median	0.075294118
Mode	0.08[b]
Standard Deviation	0.083111048
Sample Variance	0.006907446
Kurtosis	4.146363299
Skewness	1.66322481
Range	0.732903226
Minimum	0.007096774
Maximum	0.74[c]
Sum	1797.563033
Count	15126
Largest (1)	0.74
Smallest (1)	0.007096774

[a] Average Ccents/Mminute.
[b] Most common price per minute.
[c] Highest per-minute charge.

**Exhibit 13. Descriptive Statistics
for Call Duration**

Mean	4.659067169
Standard Error	0.20483191
Median	1.25
Mode	0.5
Standard Deviation	25.19182686
Sample Variance	634.6281406
Kurtosis	1058.880179
Skewness	25.32919356
Range	1517.7
Minimum	0.5
Maximum[a]	1518.2
Sum	70473.05
Count	15126
Largest (1)	1518.2
Smallest (1)	0.5
Confidence Level (95.0%)	0.401495173

[a] The longest single call lasted over 25 hours
(1518 minutes/60).

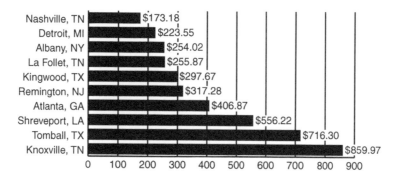

Exhibit 14. Expenses by Originating Location

- *Analyze toll-free minutes.* If some 800 or equivalent numbers have minimum charges, they should be investigated for possible cancellation. In this case, the invoice will have a charge but there will be no minutes incurred. Departments within an organization might have set up a toll-free number for a special promotion and later forgotten to cancel it. Also, those toll-free numbers showing large numbers of minutes should be called to see who answers. One St. Louis firm found that four of its largest toll-free lines charged on the invoice actually belonged to another firm. The discovery was made simply by calling to identify the answering party.
- *Determine how toll-free calls are set up to be transmitted to local premises.* For example, AT&T has two well-known toll-free services: Readyline 800 and Megacom 800. As shown in Exhibit 15, Megacom is significantly less expensive than Readyline on a per-hour basis. However, Megacom requires a dedicated circuit whereas Readyline can be installed in a switched environment (e.g., at an individual's home). Based on usage and a breakeven calculation, the organization should choose between the two services. Other carriers will have similar offerings.
- *Determine if toll-free numbers are being used where local service is available.* While this calling pattern cannot reasonably be blocked for paying customers and other outside callers, it makes sense to block employees from using the toll-free lines when a local call would provide the same service. This is typically found when traveling employees use the same 800 number to dial a local server that they use on the road.
- *Review invoices for all circuits.* Compare the locations on the invoice to an up-to-date company directory. Research any circuits that terminate in closed offices or plants. Also review usage if the business activity at those locations has been significantly curtailed. For example, a dedicated access circuit may no longer be justified and it should revert to switched access.
- *Summarize calling card transactions by terminating number.* Based on the terminating numbers for the organization's office buildings and plants, determine the cost that employees incur when calling into company premises using calling cards. Compare this to similar toll-free service to determine if a toll-free number should be used for employee calls to the office or plant.

Exhibit 15.   Tariff Pricing of AT&T Megacom 800 and Readyline 800 Services

AT&T COMMUNICATIONS
Adm. Rates and Tariffs
Bridgewater, NJ 08806

TARIFF F.C.C. NO. 2
5th Revised Rate Table 6.3.2-1
Cancels 4th Revised Rate Table 6.3.2-1
Page 1

Issued: March 30, 2001

Effective: April 1, 2001

Rate Table 6.3.2-1
AT&T MEGACOM 800 SERVICE

Service Area	Per Hour of Use					
	Business Day		Evening		Night/Weekend	
1	$21.24	I	$17.28	I	$17.28	I
2 - 6	$21.24	I	$17.28	I	$17.28	I

Rate Table 6.4.2-1
AT&T 800 READYLINE

Service Area	Per Hour of Use					
	Business Day		Evening		Night/Weekend	
1	$30.96	I	$24.48	I	$24.48	I
2 - 6	$30.96	I	$24.48	I	$24.48	I

Note: Most contracts for larger firms will provide for some percentage off the above, full-tariff rates.

Courtesy of AT&T.

- *Review invoices for T1 circuits and group by location.* Obtain current pricing for T3s (if not already in the contract) and determine if any location is near the eight-to-ten T1s versus one T3 breakeven point.
- *Obtain vendor payments in electronic form from Accounts Payable.* Ignore payments below some reasonable cutoff amount. Next, summarize by vendor and identify by visual inspection those that are likely to be telecom related. Note any payments to carriers that are not part of the negotiated plan. For example, if MCI is the carrier of choice, note any payments to AT&T, etc. These payments may represent smaller offices that have made separate, but uneconomic arrangements for telecommunications services; they are likely receiving "mom-and-pop" rates.

Local Telecom Costs

Local telecom billing is often even more convoluted than its long-distance counterpart. Understanding the structure and nuances of these bills is essential to the audit of local telephone expenses. Following are illustrations of typical LEC bills and the key information that should be extracted for analysis. When ordering copies of bills, customer service records, and other carrier information, it is important to get what is requested — avoid summary substitutes that may not provide sufficient detail.

Exhibit 16 shows an example of a LEC billing summary page. Note in Exhibit 17 that recurring charges are not detailed on the LEC invoice. The billing detailed charges are shown in Exhibit 18, and Exhibit 19 illustrates call detail for a specific local line (i.e., 517-337-2445). Note the total usage for individual lines for the billing cycle.

The weakness of the regular LEC bill is that it does not list any detail for recurring charges. Because recurring charges are often where the errors are found, a new document, the CSR (Customer Service Record) is required. Example pages of a CSR are shown in Exhibits 20 and 21.

Within the customer service record, USOCs (universal service order codes) are key to understanding the plethora of charges heaped upon the customer. Exhibit 22 shows some typical USOCs. There are thousands, although most customers will find fewer than a hundred on a typical bill. One of the telecom dreams that was, unfortunately, never realized was the "universal" in USOC. In reality, these codes vary by supplier.

Exhibit 23 shows details for the itemized line charges. The sample USOC value listed in Exhibit 22 can be used to determine that the 1MB charge is for a business message rate line. Note that the USOC codes listed are only a fraction of USOC codes in use by the local carriers.

Each physical line has its own listing on the customer service record. The number of charges for a particular service, such as 9ZR, should not exceed the number of lines. Exhibit 24 shows two lines on the CSR.

In addition to the above, a trunking worksheet (developed in part from PBX reports) and a worksheet for equipment may need to be developed. See Appendix H for more details.

Exhibit 16.   LEC Billing Summary Page

AMERITECH   Account Number
Bill Payment Center   517 337-2445 3021
Saginaw, MI 49784-0003

PRICEWATERHOUSECOOPERS   March 22, 1999
2001 Ross Avenue
E Lansing, MI 48820

Amount		Amount	
If Paid After		If Paid on or before	
May 12, 1999	309.17	May 12, 1999	306.88

Detach and mail top section with your payable to Ameritech. Write account number on check.
PLEASE ALLOW FIVE DAYS WHEN PAYING BY MAIL OR AT AUTHORIZED AGENCY.
Mail payments to: Ameritech, Bill Payment Center, Saginaw, MI 49784-0003

                                                      517 337-2445 3021

CUSTOM BUSINESS SERVICES
BILLING SUMMARY   **MAR 22, 2001**

	Payments	Adjustments	Balance	Current	Total
Previous Bill	Thank You			Charges	Amount Due
292.36	292.36	0.00	0.00	306.88	306.88

(Second Page)
SUMMARY OF CURRENT CHARGES

Ameritech Local Service
For Detailed Charges See Page 2

Monthly Service	240.84
Local Calls	18.19
Information Charges...	.00
Long Distance..	7.83
Local, State and Federal Charges	14.76
Taxes (Fed 7.80)(St 15.59)	23.39
Ameritech Local Service CURRENT CHARGES...............................	**305.01**

AT&T
For Detailed charges See Page 41

Information Charges...	.75
Long Distance..	.97
Taxes (Fed .05)(St .10)	.15
AT&T CURRENT CHARGES..	1.87
TOTAL CURRENT CHARGES	**306.88**

Exhibit 17.   LEC Summary Bill: Summary of Current Charges

SUMMARY OF CURRENT CHARGES

Ameritech Local Service
For Detailed Charges See Page 2
Monthly Service 240.84

Exhibit 18.   Billing Detailed Charges

DETAILED CHARGES

See Page 2 for Customer Service Telephone Numbers

Page 3
ACIS

Long Distance — continued

No.	Date	Time	Place Called		Number	Code	Min	
1	3-16	1101A	WEBBERVILLE	MI	517 555-4899	D	2	.50
			Total Calls Charged to 517 337-2225..					3.10
			Calls Charged to 517 337-2445					
2	3-02	1158A	CHARLOTTE	MI	517 543-4333	D	3	.74
3	3-02	1218P	JACKSON	MI	517 784-5680	D	2	.53
4	3-08	225P	JACKSON	MI	517 769-6613	D	1	.27
			Total Calls Charged to 517 337-2445..					1.54
			Calls Charged to 517 337-2850					
5	3-19	301P	CHARLOTTE	MI	517 543-6750	D	1	.25
			Total Calls Charged to 517 337-2850..					.25
			Calls Charged to 517 337-4820					
6	3-01	443P	JACKSON	MI	517 784-5680	D	2	.53
7	3-10	916A	JACKSON	MI	517 769-6613	D	1	.27
8	3-10	919A	CHARLOTTE	MI	517 543-4333	D	1	.25
			Total Itemized Calls					1.05
			Calls to AIRTOUCH CELLULAR NETWORK					
9	3-18	345P	MOBILE USE	CH	517 290-4989	D	1	.39
			Total Calls to AIRTOUCH CELLULAR NETWORK					.39
			Total Calls Charged to 517 337-4820..					1.44
			Calls Charged to 517 337-4825					
10	3-15	1004A	WEBBERVILLE	MI	517 521-4379	D	5	1.23
			Total Calls Charged to 517 337-4825..					1.23
			Calls Charged to 517 337-4827					
11	3-15	1059A	JACKSON	MI	517 784-2856	D	1	.27
			Total Calls Charged to 517 337-4827..					.27
			Total Long Distance Charges					7.83

Local, State and Federal Charges

Emergency 911 Service	2.88
Emergency 911 Operational Assessment...	6.96
Number Portability Surcharge..	4.92
Total Local, State and Federal Charges..	14.76

Taxes

Federal at 3%	7.80
State at 6%	15.59
TOTAL Ameritech Local Service CURRENT CHARGES..	305.01

FOR CALLING CODES PLEASE SEE THE BACK OF THE FIRST PAGE

Exhibit 19.   Call Detail for a Specific Local Line

			Calls Charged to 517 337-2445					
2	3-02	1158A	CHARLOTTE	MI	517 543-4333	D	3	.74
3	3-02	1218P	JACKSON	MI	517 784-5680	D	2	.53
4	3-08	225P	JACKSON	MI	517 769-6613	D	1	.27
			Total Calls Charged to 517 337-2445..					1.54

Exhibit 20. Customer Service Record Example Page 1

BL GRP	CODE & QNTY	DESCRIPTION	Unit RATE	TOTAL	TAX FSCMXT
	ZBU	CB, MNF-FMO			
	ZCPI	U			
		-MAIN LISTING			
	LN	P*W*C			
	LA	2001 ROSS AVENUE, E LANSING			
	LOC	(DES SUITE 250)			
	SIC	6141			
		-BILLING INFORMATION			
	BN1	PRICEWATERHOUSECOOPERS			
	BA1	2001 ROSS AVENUE			
	PO	E LANSING MI 48820			
	ZGC	230000000			
	SS	000-00-0000Y			
	TAR	NONE			
	ZCPI	U, SYSTEM, 9-14-96			
		-SERVICE AND EQUIPMENT			
	HTG	A 2445, 4820, 4825, 4827			
	1	1MB / PIC ATI / PCA BO, 07-12-096 / ZPIC A13 / LPCA BO, 07-12-96 / LCC 1M9 / HTG A	13.20	13.20	TTNTNN
	1	SCFXE	0.02	0.02	TTNTNN
	1	TTB	0.00	0.00	NNNNNN
	1	RTV1N	0.00	0.00	TTNTNN
	1	UXWAH	0.58	0.58	NNNNNN
	1	UXTAH	0.24	0.24	TTNTNN
	1	9ZR	5.62	5.62	TTNTNN
	1	NSR	0.41	0.41	TTNTNN
	1	1MB / TN 337-2225 / PIC ATI / PCA BO, 07-12-96 / 7ZIC A13 / LPCA BO, 07-12-96 / LCC 1M9	13.20	13.20	TTNTNN
	1	SCFXE / TN 337-2225	0.02	0.02	TTNTNN
	1	TTB / TN 337-2225	0.00	0.00	NNNNNN
	1	RTV1N / TN 337-2225	0.00	0.00	TTNTNN
	1	UXWAH / TN 337-2225	0.58	0.58	NNNNNN
	1	UXTAH / TN 337-2225	0.24	0.24	TTNTNN
	1	9ZR / TN 337-2225	5.62	5.62	TTNTNN
	1	NSR / TN 337-2225	0.41	0.41	TTNTNN
	1	1MB / TN 337-2225 / PIC ATI / PCA BO, 07-12-96 / ZPIC A13 / LPCA BO, 07-12-96 / LCC 1M9	13.20	13.20	TTNTNN
	1	SCFXE / TN 337-2531	0.02	0.02	TTNTNN
	1	TTB / TN 337-2531	0.00	0.00	NNNNNN

Having developed an understanding of LEC billing, you can start the analysis. Steps include:

1. Create a database
2. Populate a database
3. Review for missing items
4. Put it all together
5. Review common problem areas
6. Use intuition during investigation

Exhibit 21. Customer Service Record Example Page 2

```
|                        CUSTOMER SERVICE RECORD                                          |
|                                                                                         |
| BL   | CODE & |                                        | Unit  |        | TAX         |
| GRP  | QNTY   |              DESCRIPTION               | RATE  | TOTAL  | FSCMXT      |
|      |        |                                        |       |        |             |
|      |     1  | TTB      / TN 337-4871                 | 0.00  |  0.00  | NNNNNN      |
|      |     1  | RTV1N    / TN 337-4871                 | 0.00  |  0.00  | TTNTNN      |
|      |     1  | UXWAH    / TN 337-4871                 | 0.58  |  0.58  | NNNNNN      |
|      |     1  | UXTAH    / TN 337-4871                 | 0.24  |  0.24  | TTNTNN      |
|      |     1  | 9ZR      / TN 337-4871                 | 5.62  |  5.62  | TTNTNN      |
|      |     1  | NSR      / TN 337-4871                 | 0.41  |  0.41  | TTNTNN      |
|      |        |                                        |       |        |             |
|      |        |      TOTAL EXCLUDING TAXES             |       | 240.84 |             |
|      |        |                                        |       |        |             |
|      |        | AMOUNT SUBJECT TO FEDERAL TAX          |       | 233.88 |             |
|      |        |   AMOUNT SUBJECT TO STATE TAX          |       | 233.88 |             |
|      |        | AMOUNT SUBJECT TO MUNICIPAL TAX        |       | 233.88 |             |
|      |        |                                        |       |        |             |
|      |        |                                        |       |        |             |
|      |        | -SERV & EQUIP ACCOUNT SUMMARY          |       |        |             |
|      |    12  | NSR      Number Portability Surcharge  |       |        |             |
|      |    12  | RTV1N    900/976 Blocking              |       |        |             |
|      |    12  | SCFXE    AETCP Offset                  |       |        |             |
|      |    12  | TTB      TouchTone Service             |       |        |             |
|      |    12  | UXTAH    Ingraham County E911          |       |        |             |
|      |    12  | UXWAH    Ingraham County Surcharge     |       |        |             |
|      |    12  | 1MB      Individual Measured Business  |       |        |             |
|      |    12  | 9ZR      Federal Access Charge         |       |        |             |
```

7. Examine invoices for premium services
8. Evaluate invoices for use of switched traffic

A sample analysis database is shown in Exhibit 25. This will vary, depending on the LEC, services used, etc. Population of the database includes the following steps:

1. Enter the customer-defined unique identifier:
    a. Name, number, city, etc.
2. Enter the main listed phone number, which is the primary telephone number.
3. Enter the physical location of the site and the equipment.
4. Enter the site-specific client contact.
5. Enter the name of the local exchange provider, along with the contact name and number.
6. Enter the LEC account number(s) for this location.
7. Enter dates that the service records were requested and received.
8. Enter dates that the current invoice was requested and received.
9. Enter the actual line numbers from the CSR into the database.
10. Enter IXC provider name and contact information.
11. Continue to enter this information for all IXCs.
12. Enter account numbers for voice services.
13. Enter circuit numbers billed to voice invoices.
14. Enter account numbers for data services.

Exhibit 22.   Example Universal Service Order Codes (USOCs) Found in CSR

113	Regionserve Business Indi+, 2 Way
113CL	Regionserve-Bus Indiv+W/Caller ID
1FB	Flat Rate Line, Business, Two-Way
1FBCL	Flat Rate Business Line, Caller ID
1LD1E	Primary Rate ISDN Non-Dis, Month to Month
1MB	Message Rate Line, Business
1MBC1	Measure Rate Business Line With Caller ID
1MBGE	Message Rate Line, Business, Georgia Community Calling
1MG	Business Measured Rate Line Business, Two Way, Non Hunting
1MGCL	Measure Rate Business Line With Caller ID
1MH	Measure Rate Business Line With Caller ID
1S8.	Business Line, Economy Service, Local Option Service Rate, Two Way
1S8CL	Business Line, Economy Service, Local Option Service Rate, Two Way, With Caller ID
1ZJ	Business Line Standard Local Operational Service Rate, Business, Std Option, Two Way
1ZJCL	Business Line Standard Local Operational Service Rate, Business, Std. Option, Two Way
3QM1X	Essx Nar Region serv, Msd Rate Inward Only
3QMCX	Essx Nar Region serv, Msd Rate Both Way
3QMOX	Essx Nar Region serv, Msd Rate Outward Only
7FB	Business Line — Auxiliary Line
7FBCL	Business Line — Auxiliary Line With Caller ID
ACB	Area Calling Service Business, Expanded Local And Calling, Economy Option
ACBCL	Area Calling Service Business, Expanded Local And Calling, Economy Option With Caller ID
ASB	Area Calling Service, Expanded Local Area Calling Service, Std. Option
ASBCL	Expanded Local Area Calling Service, Standard Option With Caller ID
B1M	Business Line — Measured
B1MCL	Business Line — Measured
B2K1D	Flat Rate Usage Charge, Business Inward (W/Lud)
B2K1K	Flat Rate Usage Charge, Business Inward
B2K2D	Flat Rate Usage Charge, Business Both Way (W/Lud)
B2K2K	Flat Rate Usage Charge, Business Both Way (W/O Lud)
B2K2P	Flat Rate Usage Charge, Business Both Way, Premium Calling
B6P	Area Plus Service Business
B6PCL	Area Plus Service Business With Caller ID
BC1	Business Choice
BC2	Business Choice
BD1	Business Plus Service, Florida, Option 1, Flat Rated Plan

Note: Includes only a small subset.

Exhibit 23. Details for Itemized Line Charges

			-SERVICE AND EQUIPMENT				
	HTG		A 2445, 4820, 4825, 4827				
		1	1MB[a] / PIC ATI / PCA BO, 07-12-096 / ZPIC A13 / LPCA BO, 07-12-96 / LCC 1M9 / HTG A	13.20	13.20	TTNTNN	
		1	SCFXE	0.02	0.02	TTNTNN	
		1	TTB	0.00	0.00	NNNNNN	
		1	RTV1N	0.00	0.00	TTNTNN	
		1	UXWAH	0.58	0.58	NNNNNN	
		1	UXTAH	0.24	0.24	TTNTNN	
		1	9ZR	5.62	5.62	TTNTNN	
		1	NSR	0.41	0.41	TTNTNN	

[a] 1MB means Message Rate Line, Business.

Exhibit 24. USOCs Shown Separately for Each Line

		1	1MB / PIC ATI / PCA BO, 07-12-096 / ZPIC A13 / LPCA BO, 07-12-96 / LCC 1M9 / HTG A	13.20	13.20	TTNTNN	
		1	SCFXE	0.02	0.02	TTNTNN	
		1	TTB	0.00	0.00	NNNNNN	
		1	RTV1N	0.00	0.00	TTNTNN	
		1	UXWAH	0.58	0.58	NNNNNN	
		1	UXTAH	0.24	0.24	TTNTNN	
		1	9ZR	5.62	5.62	TTNTNN	
		1	NSR	0.41	0.41	TTNTNN	
		1	1MB / TN 337-2225 / PIC ATI / PCA BO, 07-12-96 / 7ZIC A13 / LPCA BO, 07-12-96 / LCC 1M9	13.20	13.20	TTNTNN	

Note: Each line has its own set of USOCs.

15. Enter circuit numbers billed to data invoices.
16. Enter equipment type and corresponding vendor information. For example, equipment type can be key system, hybrid, or PBX. Equipment vendor can be Avaya, Nortel, Siemens, or 3COM.
17. Review your worksheet for any missing items. For example, each local line number should have an entry on the CSR and each line on the CSR should have an entry on the worksheet.

When the analysis is complete, the database/worksheet should look something like Exhibit 26.

A telecommunications audit is more unwieldy than it is complex. There are many possible variations and we cannot address the myriad of potential findings in one book. Appendix H lists worksheet content for most telecom cost elements. These should be considered when performing fieldwork. The following illustrations are typical of findings resulting from this level of analysis.

Exhibit 25.   Elements of an LEC Analysis Database

Branch number or location identifier
Main number
Street address
City, state, zip
Contact
Contact telephone number
LEC: local exchange provider name
LEC contact name
LEC contact telephone number
LEC account number
CSR requested
CSR received
Current invoice requested
Current invoice received
Local lines itemized, fax and modem lines noted
IXC provider
IXC contact name
IXC contact telephone number
Voice account number
Circuit numbers
Frame or data account number
Circuit numbers
Equipment type
Equipment vendor
Vendor contact name
Vendor contact telephone number

Findings

Exhibit 27 illustrates a findings sheet from a telecom billing audit. A detailed summary spreadsheet is shown in Exhibit 28. In summary, the findings in the above review fell under the following categories. Exhibit 29 presents savings in graphical format.

- Non-essential charges:
  - Premium charges
  - Remote call forwarding
  - Premises wire maintenance plan
  - Off-premise extensions
  - WATS saver plan
  - Calls to cellular service
- Unnecessary circuits:
  - Local lines
  - ISDN BRI

Exhibit 26.   Cost Savings Database/Worksheet

LEC	IXC	Termination	Usage AT&T	Usage LEC	LEC Calls	LEC Minutes	AT&T Calls	AT&T Minutes
					216			
NA	NA		$123.96	NA			1092	2714.8
√	√	Legend		$1.54	6	3		
√	√	Legend	$2.60	$0.27	1	1	29	33.6
√	√	Legend	$6.60	$1.23	5	5	21	47.8
√	√	Legend	$7.91	$1.44	5	4	26	50.5
√	√	Stand-alone						
√	√	Stand-alone		$0.25				
√	√	Stand-alone	$49.37	$3.10			382	898.8
√	√	Stand-alone		$0.25				
√	√	Stand-alone	$49.37	$3.10				
√	√	?						
√	√	?						
√	√	?						

Note: Not all fields shown.

- Other issues:
  - Potentially unauthorized charges for:
    - Services on the LEC bill
    - AT&T usage not on VTNS invoice
    - Uneconomical use of non-AT&T VTNS calling card
    - No 900/976 blocking
  - No restrictions on authorization to change local service
  - Incorrect coding of directory advertising expense
  - Nonconsolidated LEC bills

Summary

Telecom auditing, whether by in-house personnel or outside specialists, is mandatory if an organization intends to control telecom costs. The unexamined telecom bill, like the unexamined life, is worth reviewing. By simply implementing a process to review bills and supporting documentation, businesses can typically shave five to thirty percent off their telecommunications costs. In addition, a review of telecom costs will reflect the extent to which these expenses are aligned with strategic business directions.

Notes

1. Proper allocation of telecom costs by department or individual is a major issue for many organizations. For example, universities, hospitals, and hotels must account for costs based on some logical and supportable algorithm.

Exhibit 27.   Billing Audit Findings Sheet

Office 6555, East Lansing, MI

Savings and/or Recovery Summary

Issues		Monthly	Annualized
1	Uneconomical Routing: Toll Calls Routed LEC vs. AT&T Switched	$ 3.91	$ 46.92
2*	Unnecessary Circuits: Local Lines	100.35	1,204.20
3	Uneconomical Routing: Local Measured Calls Routed LEC vs. AT&T Switched	5.88	70.56
4	PIC Charges: Update VTNS Database	12.50	150.00
5	Uneconomical Routing: Calls Routed AT&T Switched vs. AT&T Dedicated	11.17	134.04
	Total Savings or Recovery	**$ 165.76**	**$1,605.72**

*Indicates that recovery may be possible with additional evidence.

Circuit Inventory

Legend:
√ Validated
? See **Cost Savings and/or Recovery** section for more detail.

Phone Number	Type	Pre-T1	Post-T1 Plan	Post-T1 LEC	Post-T1 VTNS
1 (517) 337-2445	Main	√	√	√	√
2 (517) 337-4827	Hunt	√	√	√	√
3 (517) 337-4825	Hunt	√	√	√	√
4 (517) 337-4820	Hunt	√	√	√	√
5 (517) 337-4860	Fax	√	√	√	√
6 (517) 337-2850	Fax	√	√	√	√
7 (517) 337-2225	Fax	√	√	√	√
8 (517) 337-4871	Modem	√	√	√	√
9 (517) 337-2741	Modem	√	√	√	√
10 (517) 337-2531				√	√
11 (517) 337-2625				√	√
12 (517) 337-2633				√	√

Effective Cost-Per-Minute

	Intrastate/ Intra-LATA		Intrastate/ Inter-LATA		Interstate Calls
Switched	N/A		8.0¢	√	9.3¢
Dedicated	1.9¢	√	5.7¢	√	4.5¢
On-Net	N/A		3.5¢	√	4.3¢

√ Indicates that the figure reflects contracted tariff rates.

Frame Relay Review

Item	Cost
Access Port	$ 580.00
PVC 64Kbps	56.00
Customer Network Management Service	5.00
Total (Taxes, Surcharges and Fees Excluded)	**$ 641.00**

Exhibit 28. Summary Findings: Cost Savings and Recovery Issues

No.	Service	Issue	Findings	Recommendation	Impact	CI
1	Local	Uneconomical routing: toll calls routed LEC versus AT&T switched	The March 22 bill had detail for 17 calls. These calls should have been routed over the T1. LEC toll cost was $4.48; AT&T dedicated cost will be $0.57.	Direct staff to use the AT&T T1 for all usage-sensitive calls.	$3.91 in monthly savings	
2	Local	Unnecessary circuits: local lines	The quantity of lines exceeds the predetermined amount.	Disconnect five lines. These five lines were chosen because no discernable usage was found. However, Lucent should be called to check and reprogram the equipment if necessary.	$100.35 in monthly savings	
3	Local	Uneconomical routing: local measured calls routed LEC versus AT&T switched	The March 22 bill had local charges for 216 calls. These calls should have been routed over the T1. LEC cost was $18.19; AT&T dedicated cost will be $12.31.	Direct staff to use the AT&T T1 for all usage-sensitive calls.	$5.88 in monthly savings	
4	Long distance	PIC charges: update VTNS database	Disconnect lines identified in Issue 2.	Notify AT&T to remove from the VTNS database the five local numbers to be disconnected.	$12.50 in monthly savings	
5	Long distance	Uneconomical routing: calls routed AT&T switched versus AT&T dedicated	There were 76 calls routed switched rather than over the T1. The switched costs were $17.11; the dedicated cost will be $5.94.	Direct staff to use the AT&T T1 for all usage-sensitive calls.	$11.17 in monthly savings	

Note: A checkmark in column CI would indicate that a document for the LEC or IXC had been created for the issue.
A * would indicate that recovery would be possible with additional evidence.

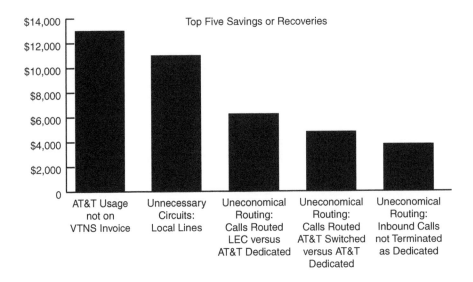

Exhibit 29.   Graphical Display of Largest Savings

Chapter 4

Wide Area Networking

The cost to transmit information to distant locales often exceeds the combined expenses for premise equipment, telecom salaries, and maintenance. Hence, there is a high payback for increased efficiency and strong controls over the design, procurement, and monitoring of the organization's wide area network. This chapter is divided into two sections: a detailed methodology for a wide area network review, and a presentation of communications technologies that can potentially reduce costs or enhance service levels.

Wide Area Network Technology Options

Frame Relay

In the 1970s and 1980s, IBM mainframes were so dominant that the comment "no one ever got fired for buying IBM" became a cliché. Frame Relay now appears to have a similar cachet — the service is low cost, almost ubiquitous in the United States and reliable. Also, contrary to general perception, Frame Relay is expandable well beyond T1 speeds and, in fact, has no specific bandwidth limit (e.g., Verizon offers speeds up to 44 Mbps). So any organization considering a WAN deployment should include Frame Relay as a priority option.

Why Frame Relay rather than traditional circuits (e.g., T1s or ISDN)? Frame Relay costs less for the same throughput because it more efficiently uses bandwidth. As the successor to the hoary X.25 standard,[1] Frame Relay allows multiple customers to share the bandwidth of a physical connection by taking advantage of the bursty nature of data transmissions (bandwidth on demand). It supports applications such as host-to-host/LAN-to-LAN links, telecommuting, multiple user Internet access, PBX-to-PBX communications, and passable voice/video communications.

The cost for Frame Relay service usually includes three elements:

1. PVC (private virtual circuit), which is usually related to the CIR (committed information rate)
2. Port charges
3. Access to the premises

It would seem that with only three major cost elements, comparing service offerings would be straightforward. Unfortunately, there are a number of factors that complicate the analysis. Following are key factors to consider.

Port Size, CIR, and Discard Eligible Flag

A rough rule of thumb that some network designers use is to set the CIR at half the port size (e.g., a PVC with a port size of 512 kb might have a CIR of 256 kbps). A better approach is to understand the bandwidth requirements of the organization's users and applications and set port size and CIR at optimum levels.

Assume, for example, a Portland field office is connected to the New York headquarters building. Portland has low bandwidth requirements but needs to be able to connect at any time (and not be subject to bottlenecks during busy times of the day). Portland might have a port speed of 128 kbps and a CIR of 64 kbps. In addition, there are six other field offices that transmit to headquarters, with the same specifications. The headquarters port speed is set at 256 kbps, with a CIR of 128 kbps. Clearly, headquarters is seriously oversubscribed. That is, if all sites transmit at once, headquarters will not be able to handle the volume. If the business environment is such that the network designer *knows* all six will not be transmitting at once, this can be a practical way to minimize costs.

If the network designer also knows that users in field offices can tolerate some transmission delay, further savings can be obtained by reducing the CIR, maybe even down to zero. At zero CIR, all packets are marked as "discard eligible" and are marked for a later transmission.

Asymmetric PVCs

Some carriers, such as AT&T, allow PVCs to be configured with CIRs (committed information rates) that are not equal in both directions. For example, assume a firm's corporate office is in Knoxville, Tennessee, and one of its field offices is in Houston, Texas. Data transmission from Houston to Knoxville may require a CIR of 64 kbps, whereas Knoxville to Houston may only require a 16-kbps CIR. If the carrier permits asymmetric PVCs, they should be considered because many times traffic is unequal between sites. Because the CIR is one factor driving Frame Relay charges, use of this technique can drive down costs with no decrease in service levels to the organization. Many WANs using Frame Relay have been implemented without fine-tuning for unequal traffic.

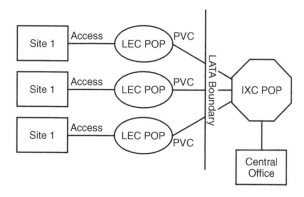

Exhibit 1.   Multi-Carrier Frame Relay Configuration

Multi-Carrier Networks

Many Frame Relay networks are single vendor from the IXC (interexchange carrier) POP to the destination. The local access link may be provided by the LEC, but the Frame Relay network itself is all one vendor. An alternative and more economical solution is to use a LEC Frame Relay network to concentrate traffic to a hub within an intraLATA area, and then transmit to major sites using IXC Frame Relay facilities. The critical factor is the access link. There are two disadvantages to this approach: (1) additional time is required to negotiate and manage separate vendors, and (2) some network management information is lost when Frame Relay packets cross vendor boundaries.

Exhibit 1 illustrates the multi-carrier approach. This solution only makes sense if the organization's topology fits the scenario — smaller locations in relatively close proximity to a hub location (within an intraLATA boundary). The alternative to this approach is to connect each site directly to the IXC POP.

PVC versus SVC

Initially, carriers set up Frame Relay circuits with dedicated, permanent virtual circuits (PVCs) that required an always-up access circuit to the POP. However, switched virtual circuits (SVCs) are now available for organizations that need (1) less frequent access to the network, or (2) more dynamic connection requirements. An SVC is started by the user, then the data is sent and the connection is torn down as in a traditional telephone call. SVCs are less expensive than PVCs up to a point (similar to traditional dial-up per-minute charges versus a dedicated circuit). Aside from lower transmission costs for limited duration sessions, SVCs offer other potential benefits:

- Reduced equipment costs (FRADs[2] and router serial ports) relative to a complete PVC implementation, particularly as the network grows in a highly meshed configuration.

- *Inexpensive disaster recovery capability.* Ongoing backup PVC costs are not incurred and regular database updates for backups can be scheduled as appropriate.
- *Temporary, any-to-any connections.* These limited-duration links eliminate the need for PVCs between sites that only occasionally communicate with each other.
- *Simplified administration.* Preconfiguring and managing PVC changes are time-consuming. For highly meshed networks, SVCs can reduce network configuration maintenance.

The above advantages are contingent on the availability of SVCs from the carrier and on user requirements. Also, at certain volumes of traffic, SVCs are no longer economical — sites should be periodically reviewed for appropriate technology. Unfortunately, many carriers do not offer SVCs.

Frame Relay over DSL

Increasingly, CLECs are offering Frame Relay via a DSL link (FRoDSL). Combined with the increased ability of providers to monitor *commercial* DSL and provide service-level guarantees, this option can provide significantly lower access costs.

Voice Communications Networking

Voicemail

Voicemail, which became widespread in the 1980s, was originally considered a substitute for a live person at the other end of the line. More recently, however, a shift in usage toward *intentional messaging* has occurred. Where there is no need for dialogue, voice messages can be recorded and sent quickly to an individual extension or distribution list.

Most major voicemail vendors have long provided the ability to transfer voicemail messages from one location to another over dedicated lines or the PSTN (public switched telephone network). For example, Avaya's Audix system can forward messages to another Audix server or to a different vendor's voicemail system using the AMIS (Audio Messaging Interchange Specification).

More recently, a new standard called VPIM (Voice Profile for Internet Messaging) has been developed, which allows voice messages to be packetized and sent over IP networks (or the public Internet). Most major voicemail vendors, including Avaya, Nortel, Siemens, and others, are implementing this standard into their voice messaging products. VPIM provides both economic and functional benefits:

- *Conserves bandwidth.* The message is packetized and compressed to one half its original size.
- *Simplifies distribution.* As more voicemail systems become VPIM compatible, distribution to multiple locations is easier.

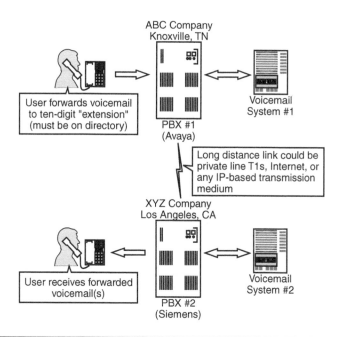

ABC Company
Knoxville, TN

User forwards voicemail to ten-digit "extension" (must be on directory)

PBX #1 (Avaya)

Voicemail System #1

Long distance link could be private line T1s, Internet, or any IP-based transmission medium

XYZ Company
Los Angeles, CA

User receives forwarded voicemail(s)

PBX #2 (Siemens)

Voicemail System #2

Exhibit 2. Transfer of Voicemail Messages Using VPIM Protocol

- *Improves efficiency of message broadcast.* The older AMIS system sent messages one at a time, even if many users at a distant location were receiving the same message. VPIM sends a single message, which is then addressed to multiple recipients, resulting in both a quicker and more efficient (i.e., less bandwidth) transmission.
- *Integrates easily with unified messaging.* Sending and receiving voicemail messages in VPIM format is more straightforward, because the transmission is treated as a special, multimedia e-mail.

Exhibit 2 illustrates the use of VPIM for voicemail message transmission.

Virtual Private Network (VPN)

The term "virtual private network" has become closely linked with substitution of an IP-based public network (usually the Internet) for dedicated or leased facilities. Instead of leasing a T1 or Frame Relay circuit to link office A to a distant office B, an encrypted "tunnel" can be established across the Internet to securely transport data packets. Originally, carriers such as AT&T used the concept of VPN (called SDN by AT&T) to describe a logical private network for each customer using the service. The term "virtual" was used because the actual hardware, software, and circuits are shared among all the carrier's customers, but the end customer perceives the service as a dedicated facility.

VPNs reduce long-distance communications costs — particularly for international sites — by eliminating much of the IXC expense. However, there are start-up and maintenance charges that can make a VPN implementation

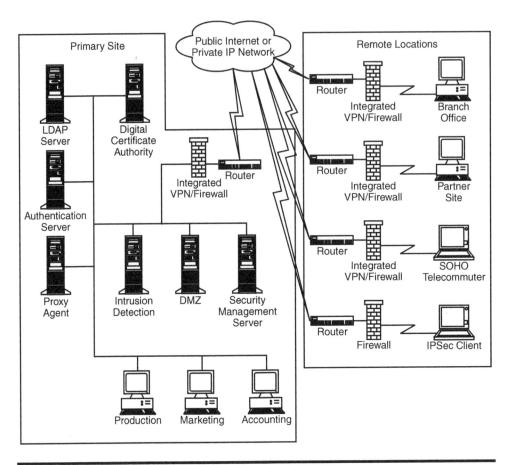

Exhibit 3.   VPN/Firewall Deployment with Security and Monitoring

uneconomical for certain volumes of traffic. Also, VPNs that use the public Internet are subject to the vagaries of events on the Net — congestion, irregular quality of service, etc.

Exhibit 3 shows a typical VPN configuration. The example shown is for data communications only. Although voice over the public Internet may yet have its day, currently the quality of service (QoS) on the Internet is not adequate for most enterprises. Voice-over-IP, using *private* transmission facilities with guaranteed QoS, is discussed in another section of this chapter.

Generally, most medium to large organizations that have multiple, dispersed sites can use VPN technology to supplement (rarely to completely eliminate) their existing wide area networks. The likelihood of a good fit increases dramatically if the organization incurs a large dialup (800 number) bill, typically associated with a RAS (remote access service) implementation. Indeed, organizations such as PricewaterhouseCoopers, having thousands of professionals on the road, have saved hundreds of thousands of dollars annually by sharply reducing long-distance dialup minutes.

When considering implementation of a VPN, there are a number of financial, business, and security issues to consider:

- Advantages:
  - Replace some dedicated lines, such as T1s, with transmission over the Internet (e.g., backup T1s could be eliminated). The organization must be aware of the caveats, such as the potential for Internet congestion and poor quality of service.
  - Eliminate some or most RAS dial-up charges. While ISPs may charge a per-hour charge for users tunneling through a VPN, those charges are significantly less than IXC per-minute charges. For example, a large organization might negotiate a $1-per-hour ISP connect time charge, whereas the same charge for an hour of toll-free dial-up could be $5.00.
  - Enable quick bandwidth increases by adding additional ports (compared to lead-times of two to eight weeks for new T1/T3 services).
  - Facilitate extranets for customers, suppliers, and partners, and provide additional E-commerce functions.
  - Make secure intranets available to field offices around the world (at a reasonable cost).
  - Provide high-speed services to telecommuters who have broadband access in the home/small office. For example, VPNs can operate over cable modem lines or DSL. With this capability, some jobs can be accomplished off site that might otherwise require office space/equipment.
  - Reduce management costs of a WAN by using a fully integrated, secure VPN solution, in contrast to the traditional plethora of network access gear.
  - Reduce the number of access lines for some field offices. If the office has a separate line for Internet access and data communications (e.g., for Frame Relay), VPN can eliminate one access line.
  - EDI (electronic data interchange) communications costs can be reduced by establishing an extranet using a VPN and eliminating use of a value-added network (VAN).

- Disadvantages/concerns:
  - VPNs are more complex to manage. Some organizations outsource the management of the VPN network.
  - VPN is not always the answer. For example, a small network with low bandwidth requirements may be better served via a Frame Relay solution (less expensive edge equipment, less maintenance).
  - The public Internet occasionally suffers congestion. Although this may someday change with the introduction of MLPS,[3] for the moment it is a significant concern for organizations that must have extremely high uptime. Some vendors offer fail-over capabilities that allow traffic to be sent over an alternative link (e.g., dial-up ISDN) if the Internet is congested.
  - The level of available VPN encryption, while certainly adequate for any domestic U.S. commercial needs, may not be available for some international traffic due to government restrictions. However, this may be

Exhibit 4.   VPN Savings Calculation

Average Annual Line Cost per Site	Intranet VPN Annual Cost	Traditional WAN Annual Cost
Central site	$19,300	$70,300
Regional site	$6,700	$30,100
Branch site	$2,600	$14,900
SOHO site	$260	$1,450
Total	$28,860	$116,750

Courtesy of Cabletron.

changing, at least for some countries. France, for example, has long required that encryption be no stronger than that afforded by a 40-bit key. Recently, the maximum permitted length has been increased to 128 bits, a considerable increase in security levels.

Example VPN Cost Worksheet

Exhibit 4 shows a VPN savings calculation (adapted from a Cabletron presentation). With a large number of dial-in users, the decision to implement VPN is almost a "no-brainer." Management expenses and costs of alternative links (Frame Relay, traditional circuits, ATM, etc.) often drive the payback calculation. Sometimes, calculated savings are less dramatic than this example — *caveat emptor* applies.

Wireless Options

George Gilder, a well-known champion of "infinite bandwidth," says the [wireless] spectrum is infinite, ubiquitous, instantaneous, and cornucopian.[4] Current industry trends certainly support his hyperbole. Wireless Internet services, low earth orbit satellite services, Bluetooth, fixed wireless, and many other permutations of communications over the air are providing services that were either not available or prohibitively expensive in the past.

Following are some of the technologies and applications that have the potential to streamline operations or to serve as less-expensive, ersatz land-lines.

VSAT and Geosynchronous Satellites

In the 1950s and 1960s, many organizations maintained private, point-to-point networks of terrestrial lines, often with multiple "drop-off points" in more remote areas. It was expensive and arduous to maintain this cobbled together network. For a network manager dealing with locations in the hundreds of

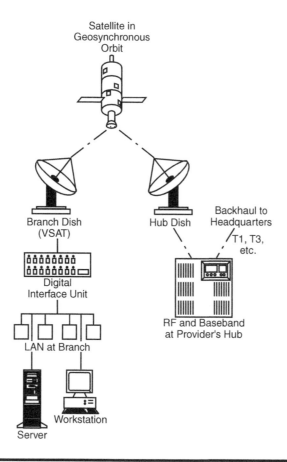

Exhibit 5. Typical VSAT Configuration

thousands, it meant maintaining contacts with many small local exchange carriers (e.g., Bob and Pat's Telephone Company) and a relatively high level of downtime (for specific sites).

As commercial satellites began to be deployed in numbers during the 1970s, the use of VSAT (very small aperture terminals) networks increased significantly. Exhibit 5 shows a simplified diagram of a VSAT network.

By using a satellite to transmit all traffic within a large geographic region (e.g., Canada, the United States, and Mexico), the need for terrestrial lines is eliminated except for the backhaul (high-capacity terrestrial circuit from the commercial satellite hub to the organization's headquarters location).

Following are advantages and disadvantages of VSAT systems versus traditional terrestrial networks:

- Advantages:
  - *Less expensive for data communications.* Satellite systems become less costly per site as the number of sites increases. The cost differential is most significant in rural areas where terrestrial access lines (e.g., for Frame Relay) are costly.

  - *Less expensive for video communications.* Terrestrial lines with the bandwidth to support video are expensive (typically requiring a minimum of 384 kbps for good quality). VSAT can deliver one-way video for a fraction of the cost of a terrestrial solution (point-to-point solutions or ATM).
  - *Known, reliable technology.* VSAT technology has been around for decades, with the dishes progressing from type I to the current type III technology. In many areas of the world, including oil rigs in the ocean, VSAT communications is the only practical alternative. Over time, techniques to optimize common protocols such as TCP/IP over higher delay satellite networks have been developed.
  - *Much quicker to deploy.* In some remote areas, the local telephone company can take months to install a data circuit. In some cases, the LEC may not be willing to incur the up-front cost. VSAT equipment, on the other hand, can be set up relatively quickly — in a week or two. The only requirements are that electrical power be available and that the satellite be within the VSAT dish line-of-sight (proper angle).
  - *Available in remote and underdeveloped areas.* VSAT technology functions well in northern Alaska, Pitcairn Island, and Tierra del Fuego.
■ Disadvantages:
  - *Only moderately high uptime.* Satellite communications cannot provide an extremely high uptime, such as 99.999. Providers such as Hughes will typically quote numbers such as 99.5 to 99.8 percent uptime. Unavoidable events such as extremely heavy rain, sunspots, and even solar transit outage[5] cause the signal to degrade and thus interrupt transmission. In some cases, interference from improperly configured ground stations (bad polarity, for example) from other carriers can weaken the signal.
  - *Transmission (propagation) delay.* Because the signal must go up 22,300 miles from the VSAT and down the same distance to the hub (ground station), there is a noticeable lag time for interactive systems (0.25 second one way, 0.5 second round trip). This reduces the popularity of VSAT for traditional voice communications, although it can work in a "take your turn ... Roger ... over" mode.
  - *High initial cost.* VSAT equipment will cost an initial $6 to $8K per site, plus any monitoring equipment that the organization chooses to use. Also, satellite contrasts are lengthy, generally five years.
  - *Limited uplink bandwidth.* While large volumes of data can typically be downloaded from the satellite (e.g., for video), uplink from a single VSAT dish is typically less than 128 kbps.
  - *Single point of failure.* Satellites have a limited life (a 15-year-old satellite is an antique) and are subject to limited fuel to keep them in proper orbit, electrical breakdowns, meteors, being hit by other satellites, and other sources of destruction. For example, PanAmSat lost its Galaxy-IV satellite in 1998, resulting in widespread loss of paging services across the United States for a few days. To mitigate this risk, organizations can obtain rights to use a backup satellite from their provider. If the primary satellite fails, VSAT dishes must be repositioned to point to the backup satellite. Repositioning can take anywhere from a few days (best case) to several weeks for a large number of sites.

An important economic consideration for an organization with a large VSAT network is hub ownership. Firms with a smaller number of sites typically use their provider's hub and receive all communications via a dedicated leased line from the provider to their headquarters site. However, even at a cost of roughly $1 million, at some point hub ownership becomes a viable option. Only organizations committed to VSAT over a relatively long time period should consider this option, because the technical staff and expertise to operate a satellite hub are considerable.

Comparison of terrestrial network costs to comparable satellite numbers depends on a number of factors, such as:

- Number of sites
- Uplink and downlink bandwidth
- Service level agreements and disaster recovery requirements
- Price of the VSAT equipment
- Maintenance costs of the equipment (dish, RF equipment)

In one recent study, the cost of supplying comparable bandwidth to 1000 sites was found to be approximately $400 to $450 per site with Frame Relay and $150 per site using VSAT services (Global VSAT Forum, *http://www.com-sys.co.uk/vsatind.htm*). The authors have seen similar figures for other firms. Including video in the mix would make the cost disparity even greater. Chapter 6 treats satellite communications in more detail.

Low Earth Orbit Satellite

The economics of wireless communications are not always suited to broad generalizations. Each business case must be considered individually. As an example, consider energy firms that use pipelines to transport natural gas across the United States. The pipelines must constantly be monitored for signs of rust to ensure that a gas rupture does not occur. One engineering technique long used by pipeline companies is cathodic protection, in which a metal rod is attached via wires to the pipelines and serves as a "sacrificial anode" to keep a correct electrochemical balance. In effect, the expendable rod rusts instead of the pipeline itself.

Because the metal rod eventually rusts out, inspections and replacements must occur on a regular basis. Trips to the more remote sites may require hours of "windshield time;" that is, a service technician driving a truck a hundred miles to spend a few minutes inspecting and possibly replacing the anode. Attaching an inexpensive transmitter to each site and then sending appropriate telemetry data to low earth orbit satellites can eliminate many of these inspection trips. The satellites, in turn, transmit to an Earth station. From there, the data is sent to a data acquisition center where appropriate maintenance reports are created.

Whether the above scenario makes economic sense depends on a number of factors: people costs, time on the road, unit capital costs (for the remote field transmitter), and system maintenance costs. Low earth orbit, or LEO,

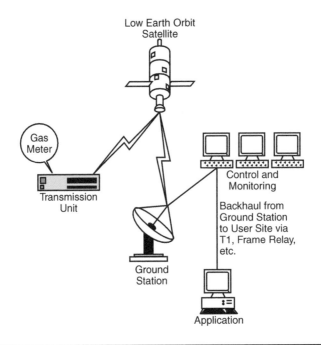

Exhibit 6.   Low Earth Orbit Transmission Example

satellite transmission is relatively expensive on a per-packet basis. However, if the application requires only a small quantity of data per month (as in the previous example), LEO technology may be a good fit. Exhibit 6, courtesy of Orbcomm and Leocell, illustrates a typical implementation.

Bluetooth

For selected environments and applications, a radio-frequency, personal area network may be superior to its wired counterpart. Bluetooth is a popular, open standard for wireless transmission over a relatively short range (10 to 100 meters). It enables functions such as:

- Wireless LAN access
- Synchronization of PDAs and laptops
- Midrange bandwidth for connection to the Internet (up to 720 kBps): any Bluetooth-enabled device, such as a mobile phone, can link to the Internet if within range of a suitable access point
- Conferencing functionality: documents and business cards can be quickly exchanged among the participants
- Faxing
- Facilitation of electronic paper transmission: for example, a sales rep could fill out a form using a Bluetooth-enabled pen that records the motion of the pen on paper and transmits the order to appropriate servers via a nearby access point or receiving PDA

To some extent, Bluetooth competes with the older Wi-Fi wireless LAN specification. However, Wi-Fi is intended as a cable system replacement and has a higher bandwidth than Bluetooth. Wi-Fi does not fill the same market space.

From a cost perspective, implementation of wireless solutions depends on the organization's workforce. Some car rental companies, for example, use CDPD (wireless) to check out returning customers. Hospitals track patient records using secure wireless technologies as well.

Fixed Wireless Broadband

The slow speed of narrowband wireless communications such as cellular voice transmissions, CDPD,[6] infrared, etc., reinforces the general perception that "wireless" denotes slow and error prone. In fact, there is no theoretical reason why fixed wireless systems cannot transmit very large quantities of data with extremely low error rates. For example, one of the networks discussed below, LMDS (local multipoint distribution services), tops out above OC-3 (155 Mbps) and is typically deployed at 45 Mbps downstream and 10 Mbps upstream. These networks use high-frequency radio connections to send and receive voice, data, and video; from the user's perspective, the result is no different than what would be expected from a copper- or fiber-based solution.

Fixed wireless solutions can often be a lower-cost alternative for broadband access, particularly in rural/low-density areas within the United States. Internationally, fixed wireless is increasingly popular due to its quick deployment, avoidance (from the carrier's perspective) of heavy infrastructure development, and, for some very poor nations, the absence of copper wires, which are sometimes stolen.

From an architectural perspective, broadband wire line access methods, such as xDSL and cable modem, compete with fixed wireless solutions. All these solutions are targeted toward solving the "last mile" problem — getting broadband to the customer's premises. When reviewing options, an organization should consider the following issues:

- ■ Advantages:
  - – Fixed wireless can be the lowest-cost alternative.
  - – The technology is quick to deploy. In some U.S. rural or international locations, wired broadband access can take several months (T1 drops can take up to nine months in some areas). Fixed wireless antennas and services can sometimes be implemented in weeks.
  - – Coverage increases are incremental (just add more receivers/transmitters).
  - – Legal/governmental regulations are much easier to address. For example, easements or special licenses are not usually required from the end customer.
  - – Under certain circumstances, fixed wireless can deliver more bandwidth than xDSL or cable.

■ Disadvantages/concerns:
  – Some technologies require line-of-sight from transmitter to receiver.
  – Tall buildings, mountains, and heavy rainfall can interfere with signals for some of the networks.
  – Standards for equipment have not yet crystallized, resulting in uncertainty in the marketplace and a smaller number of equipment vendors creating the equipment.
  – Economies of scale are still needed to achieve lowest pricing to the end user.

Listed below are the most common broadband access options using fixed wireless.

LMDS (Local Multipoint Distribution Services)

Operating in the 28-GHz range of the spectrum, LMDS provides transmission rates exceeding OC-3 (155 Mbps). A typical deployment provides 45 Mbps downstream and 10 Mbps upstream. Well suited for urban areas, LMDS can be considered "ersatz fiber" when installed with sufficient cell overlap to reduce the effects of heavy rain. One limitation is that line-of-sight is required and wireless links (transmitter and receiver) must be less than 2.5 miles from each other.

MMDS (Multichannel Multipoint Distribution Services)

Although it has been used for more than 25 years to transmit television signals, MMDS is now finding a new niche in the high-speed Internet access service world. It does not require line-of-sight transmission and can work effectively over 35 miles. At 10 Mbps, downstream speed is considerably less than LMDS but its lower frequency range makes it less susceptible to weather interference.

After gaining an understanding of the technologies available and the financial consequences of options within each technology, the next step is to perform a comprehensive review of the existing network.

Wide Area Network Review Methodology

Given the number of components and interrelated services required for a medium to large wide area network, a methodology for cost management review is essential. The methodology discussed in this section includes 11 logical steps to realize savings using network analysis and optimization techniques. These steps are best executed in parallel rather than sequentially. Each step is self-contained, including sample reports, forms, and other miscellaneous information pertinent to that step. Exhibit 7 shows a high level perspective of the entire process.

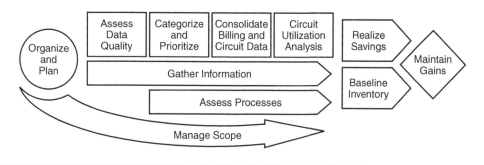

Exhibit 7. Full-Scope Cost Management Review

Steps for a full-scope cost management review include the following:

1. Organize and plan
2. Gather information
3. Assess the quality of data
4. Categorize and prioritize circuits
5. Manage scope of work changes and issues
6. Consolidate billing and circuit data
7. Initiate a circuit utilization analysis
8. Realize savings
9. Reconcile and "baseline" circuit inventory
10. Assess processes
11. Maintain the gains

Step 1: Organize and Plan

Organization and planning must occur at the project's inception to ensure that resource requirements are defined and expectations are set. Gathering information is a continuous process that is closely aligned with Steps 3, 4, 6, and 7. Managing scope and assessing processes will be performed throughout the project.

Objectives

Confirm the project scope and obtain consensus on the plan. These include the project's timeline, expected deliverables, and required resources.

Detailed Steps

- ■ Schedule a project kickoff meeting. Key staff as well as management, voice, data, and billing personnel should be included.
- ■ Focus on organization, objectives, and overall approach of the project.
- ■ Develop an initial project plan, including deliverables, timelines, and resources (see example below).

Exhibit 8. Sample Team Contact Document

Project Primary Team Contact List
As of March 5, 2002

Name	Team	Phone	Location	E-Mail
Bill Brown	Provisioning	281–358–8876	Cold Harbor	Bbrown@xyz.com
Brian DiMarsico	Project Leader	212–358–8876	New York	Bdimarsico@xyz.com
Brian Schwalen	Voice Analyst	281–358–8876	Petersburg	Bschwalen@xyz.com
Colette White	Secretary	281–358–8878	Manassas	Cwhite@xyzcorp.com
Libby Will	Billing Manager	281–358–8876	Gettysburg	Libbywill@xyzcorp.com
Bob Lee	Data Manager	713–987–6543	Cold Harbor	Rlee@xyz.com
Harish Sethuraman	Data Analyst	214–876–4654	Cold Harbor	Hsethurman@xyz.com
Janet Gunn	Tech Director	713–987–6543	Richmond	Jgunn@xyz.com
John Lerch	Voice Manager	214–876–4654	Richmond	Jlerch@xyz.com
Maggie French	Network Admin	214–876–4654	Richmond	Mfrench@xyz.com
Mike Yarberry	IS Ops & Network	713–987–6543	Montgomery	Myarberry@xyz.com
Stoney Jackson	Facilities Manager	214–876–4654	Montgomery	Sjackson@xyz.com
Lee Miller	Network Support	212–987–6543	New York	Lmiller@xyz.com
Jeb Stuart	Billing – admin.	212–876–4654	New York	Jstuart@xyz.com
Ulysses Grant	Billing – eng.	212–987–6543	New York	Ugrant@xyz.com

- Create a document that contains team members' functions, phone numbers, locations, and e-mail information (see Exhibit 8).
- Structure a weekly status report (see Exhibit 9) that includes, at a minimum, the following:
  - Activities/accomplishments
  - Issues for resolution
  - Planned activities
- Hold weekly status meetings with team members to review the status report, assign ownership for any concerns or issues identified, create a strategy to resolve issues, and summarize general project progress.
- Arrange for facility space, equipment, and computer-access privileges.

Recommendations

- The project team should list issues on the weekly status report, which may include working papers or additional information. This report is an important ground-floor communications and issue-resolution vehicle.
- Status meetings should be held weekly during the initial stage of the project, less often in later stages.
- An initial project plan should be created as soon as possible. Major activities rather than low-level details frame the project at this point.

Exhibit 10 provides an example of an initial project plan. The format and content will vary, based on the organization's project delivery standards and procedures.

Exhibit 9.   Sample Weekly Status Report

Network Inventory Analysis and Optimization Cost Initiative
Project Status — April 4, 2002

Activities/Accomplishments
1. Delivered Network Interim Analysis Report.
2. Obtained a breakdown of supplier bill group 867 for Frame Relay circuits.
3. Obtained summary (e.g., maximum, minimum, average) Web data circuit utilization information for the month of March.

Issues for Resolution
1. Without configuration, baselining, and traffic collection, NETSYS cannot be used for "what-if" analysis. Therefore, the acquisition/access to the network planning and optimization tool becomes even more crucial. Any delay in the usage of an optimization tool will delay the identification of substantial savings.

Planned Activities — Upcoming Week
1. CD-ROM access and population into the consolidated database needs to start this week.
2. Continue to collect traffic data and create a more detailed consolidated bill inventory table.
3. Start entering data into a planning tool.

Step 2: Gather Information

Objectives

Identify information required for a network analysis. The intent is to understand the level of information available, its completeness, and the quality of information available for the following:

- The network topology/connectivity
- Tools being used
- Maintenance procedures
- Linkages to other business processes (e.g., ordering, provisioning, trouble management)
- Network and site growth/build-out plans
- Business continuity requirements
- Internal traffic measurement capabilities

This understanding creates a foundation for analysis and assessment of project scope.

Process

- Conduct a quick evaluation to determine which source of information is more accurate. For example, compare the existing billing information to the network circuit inventory data. From the two sources, select the most accurate and timely source of information as the starting point for reconciliation with the other.

Exhibit 10. Sample Initial Project Plan

ID	Task Name	Duration (days)	Start	Finish	Predecessors
1	**1 Kick-off org. work**	5	Mar 03	Mar 07	
2	1.1 ID sites/contacts	2	Mar 03	Mar 04	
3	1.2 ID customer sites/contacts	2	Mar 03	Mar 04	
4	1.3 ID network plan tool in hand	1	Mar 03	Mar 03	
5	1.4 ID all carriers	1	Mar 03	Mar 03	
6	1.5 ID all contacts — local/remote	2	Mar 03	Mar 04	
7	1.6 Prioritize sites and bus functions	5	Mar 03	Mar 07	
8	1.7 Categorize market info	2	Mar 04	Mar 05	
9	1.8 Joint team meeting	1	Mar 05	Mar 05	
10	**2 Data gathering/analysis**	18	Mar 04	Mar 27	
11	2.1 Current circuit info in hand	2	Mar 04	Mar 05	
12	2.2 Current PBX info in hand	2	Mar 04	Mar 05	
13	2.3 Current bill data in hand	3	Mar 04	Mar 06	
14	2.4 Carrier contact list	2	Mar 05	Mar 06	
15	2.5 Obtain WS and CLC log-in	1	Mar 06	Mar 06	
16	2.6 Understand circuit order process	2	Mar 07	Mar 10	
17	2.7 Network diagrams	3	Mar 10	Mar 12	
18	2.8 Dial entries/modem connections	2	Mar 13	Mar 14	17
19	2.9 Hdw/switch connectivity info	4	Mar 17	Mar 20	18
20	2.10 ID current site BU pipes	2	Mar 13	Mar 14	17
21	2.11 ID replicated equipment	5	Mar 13	Mar 19	17
22	2.12 Other biller info	3	Mar 18	Mar 20	
23	2.13 Separate cust from internal ntwk	2	Mar 17	Mar 18	
24	2.14 Determine circuit field data required	3	Mar 13	Mar 17	
25	2.15 Evaluate existing diversity	5	Mar 13	Mar 19	
26	2.16 Business continuity evaluation	2	Mar 14	Mar 17	
27	2.17 Reconcile provisioning data	5	Mar 17	Mar 21	
28	2.18 Reconcile bill and provisioning	5	Mar 21	Mar 27	22
29	**3 Deliverables**	11	Mar 21	Apr 04	
30	3.1 Draft circuit utilization	5	Mar 21	Mar 27	22
31	3.2 Draft circuit discrepancy document	5	Mar 24	Mar 28	27
32	3.3 Draft cost discrepancy document	5	Mar 28	Apr 03	22,28
33	3.4 Draft baseline inventory	5	Mar 21	Mar 27	
34	3.5 Draft recommendation document	10	Mar 21	Apr 03	
35	3.6 Initial review and recommendations	1	Apr 04	Apr 04	34

■ Obtain as much BTN (bill to number)- and circuit-level detail as possible. This will increase the percentage of reconciled records between the billing and circuit inventory data and the amount of detail required for a baseline inventory.

■ Collect the following types of information to gain an understanding of the existing network (electronic format preferred):
 – Business requirements and plans
 – Service level agreements (SLAs)

  - PBX configuration and usage reports
  - Network diagrams (with circuit ID information)
  - Circuit- and BTN-level detail information
  - Telco CSRs (Customer Service Records)
  - Site and network build, move, and site closure plans
  - Web site router configuration, circuit utilization statistics, and tools/access approvals
  - Consolidated and detailed billing records, as well as general ledger information
  - Past and current order request forms
  - Critical site and application data
  - Carrier sales and provisioning contact list
- Develop a standard form that allows for the documentation of a specific information request. In addition to creating an assigned action item, it also serves as a tracking mechanism for follow-up.

In practice, the project manager should allow for the following in the time budget:

- Information must be gathered continually during the project life cycle.
- Existing telco billing information is usually more accurate than the firm's network circuit inventory.
- New information requests are generated as anomalies become apparent.
- Responses to information requests may not be a telco priority or may require considerable effort to obtain. The team must be tenacious.
- IT organizations tend to consolidate billing invoices in an effort to reduce the number of bills, through single account/summary invoices. The intent is to better manage invoice payment and reduce circuit costs because the LECs (local exchange carriers) eliminate certain charges when using summary billing. Unfortunately, this practice often precludes the development of a detailed circuit baseline inventory.

Exhibit 11 shows a sample information request form used to develop a network inventory.

Step 3: Assess the Quality of Data

Objectives

Scrutinize the data for integrity, quality, and timeliness. This provides an estimate of the effort required to correct data deficiencies in the existing network inventory database schema.

Process

- Obtain and import a copy of the existing network circuit inventory database into the preferred database (e.g., MS Access, Foxpro, or Excel).

Exhibit 11. Sample Information Request Form

Network Inventory Analysis and Optimization Cost Initiative
Tracking Request Form — March 4

Item	Information
Request 2	
Name	Call detail records/other bill data
Description	Power Bill/Billing Edge/Bill Manager info from Verizon, AT&T, MCI — also other LEC/LD consolidated billing info
Contact/source	Robert Robinson
Format	CD-ROM, paper
File type	*.dbf
Timeliness	As of last billing cycle
Requested date	3/4/02
Delivery date	
Status	As of 3/5: Robert has all other CD-ROMs but not from Verizon. Could use this info to compare against invoices and network data (existing database).

- Gather any documentation on the database schema and the logical data model for all table and element relationships. This helps to better understand the level of relationship complexity and the database usage characteristics.
- Create SQL[7] scripts to query and display various data element values. Key field element value sets, other than primary indices, that will be used for billing reconciliation should be evaluated for data quality as well as for referential integrity (see Exhibit 12 for an example of corrupt field values).
- Determine the approach to be used. Identify any steps necessary to resolve data integrity or data corruption deficiencies.
- Obtain an electronic copy and install the most current and complete version of the billing information.

Those who have participated in WAN cost management studies quickly learn the realities of billing data. For example:

- *Network analysis is hampered by inventory errors.* When key field data such as circuit ID is incorrect, the reconciliation effort is increased. Location information, circuit type, carrier, and local account billing information are additional examples of field data that are used in any reconciliation effort. Exhibits 12 and 13 show examples of corrupted field inventory data.
- *CD-ROMs from the telcos look "official" but may in fact contain errors.* For example, they tend to maintain records of circuits that have been discontinued as well as charges and credits identified by summary account (or even AT&T's MCN and Bill Group) without pertinent BTN and circuit data identified on the records. Also, the circuit inventory and circuit location tables show several records for the same circuit, due to multiple locations or number of outstanding service orders.

Exhibit 12. Sample Corrupt Field Values (Example 1)

City Name	Count
	619
???	2
???????	1
???????????	3
?????????????	1
FRANKLIN	1
FRANKLIN LADE[a]	1
FRANKLIN LAKES[a]	2
FRANKLIN LAKES,[a]	1
FRANKLIN LKS[a]	1
FRANKLYN LK[a]	1
FRNKLIN LAKES[a]	2
HOWARD BEACH	1
HUNTINGTON	6
HUNTINGTON STA	1
HUNTINGTON STA.	2
HUNTINGTON,	1
HUNTINGTON, NY	1
HUNTINGTONBCH	1
HUNTINTON	1
KNOX	1
KNOXVILLE	1
KNOXVL	1
L.I.C	2
L.I.C.	42
MORRIS PLAINS	2
MORRIS PLALINS	1
MORRIS PLNS	44
MORRIS PLNS.	1
N. Y.	1
N.Y.	24
N.Y., NY	1
N.Y., NY	2
N.Y.C	2
N.Y.C.	2
NEW YORK	4
NEW YORK CITY	1
NYC, NY	2
NYU	1

In this partial listing above, the city field has two problems. First, there are 619 records with a blank city name. Second, various spellings of the same city (e.g., Franklin Lakes") make reconciliation of circuit records more difficult.

[a] Note that "Franklin Lakes" occurs under multiple names.

Exhibit 13.   Sample Corrupt Field Values (Example 2)

Network Provider	Count
	569
614–798–33	2
AT&T	185
AT&T B-800	90
AT&T, B-80	1
ATT	181
BA, AT&T	1
BELATL	16
BELL ATL	2
BELL ATLAN	2
BELLATL/PA	42
BELSTL	1
BUSNETNY	2
N/A	4
NENEX	1
NEWPORT TE	1
NJ	1
NJBEL	15
NJBELL	1276
NJBELL\	1
NJBRLL	1
NJDELL	1
NYNEX	750
NYNEX (NE)	2
NYNEX, AT&	2
NYNEXD	1
PAC BELL	1
RGT	2
SNET	9
SPRINT	6
STRINT	1
UNITED OF	1
VOICE COM	1
WARWICKTEL	1
WILTEL	1

In this example, there are 569 blank or null rows and multiple values for the same carrier name. As a result, it may be difficult to identify the telco providing the service (unless the circuit ID value and carrier naming convention are known).

■ *Every organization that implements a circuit inventory database implements slightly different logical relationships between data elements.* These
relationships are important to understand when creating a baseline inventory from the existing database system.

Step 4: Categorize and Prioritize Circuits

Objectives

Define and prioritize categories of circuits to simplify the reconciliation and savings calculations. First, the total circuit inventory is segregated into manageable categories that can be reviewed according to business priorities. Second, the results are used as a framework or set of guidelines during evaluation of network savings.

Process

- Create an initial straw-man of categories based on several characteristics, such as:
  - Location
  - Business purpose or application needs
  - Carrier
  - Service or circuit type
- Review the straw-man with management and operations experts to confirm completeness.
- Define and create any additional categories based on current and recent business activities. Examples may include site closings, consolidations, and telecommuter programs.
- Determine business function and application priorities (e.g., backup requirements) to be used in any assessment of potential circuit restructuring or cancellation candidates.
- Identify volumes and occurrences of business activities that should be considered in potential savings assessments.
- Update any worksheets with the priority categories.

Step 5: Manage Scope of Work Changes and Issues

Objectives

Develop guidelines to manage any increased scope of work or unforeseen events that may cause the project to change direction.

Process

- Informally identify potential project roadblocks via one-on-one sessions. To further clarify objectives, hold additional meetings to explain the project goals and solicit concerns.
- Obtain funding with a cushion of 20 to 40 percent in excess of original estimates.
- Set aside additional staff resources to serve as backup to primary project team members.

Step 6: Consolidate Billing and Circuit Data

Objectives

Use the existing billing and network inventory information to create consolidated database tables for the reconciliation step.

Process

■ Create new database tables and populate them by category based on the information obtained during the data-gathering effort (e.g., billing spreadsheet data, circuit inventory data, and CD-ROM information).

■ Create a data entry tool (e.g., MS Access forms) for any information that is not in electronic form, such as hard-copy bill data.

■ Decompose the summary or consolidated billing information down to the BTN and circuit ID levels wherever possible.

■ Capture the bill amounts at the BTN and circuit levels for as many records as possible under each summary account. If this is not possible (i.e., zero bills at the BTN level), keep the summary account information in the table with the total dollar amount. In those cases where individual bill values exist, adjust the total dollars record for the individual records under that summary account.

■ Normalize the billing and circuit ID data for reconciliation purposes in a later step. For example, where there are BTN field values of 201 539-2222 in billing data and 201 539-2222 in circuit inventory, these records will never match unless a hyphen is inserted between 201 and 539, or taken out of the other field values being compared.

General Comments

■ The need to "normalize," "clean," or "scrub" the field data is more common than normally anticipated.

■ Once the database is evaluated and normalized for key comparison data, the information pertaining to ordering information must be analyzed. Order numbers and install dates provide additional information for reconciliation purposes, allowing the ability to further differentiate between two circuits that are similar in description.

Sample Consolidated Account Number Detail

Exhibit 14 illustrates the expansion of summary records. It starts with a single Bell Atlantic consolidated account number, 201 X52-0066-999 32, and expands the information available. Originally, only total dollars were available. By obtaining hardcopy billing records, performing a review, and entering additional data into a spreadsheet, much useful detail (such as Bill Number and circuit information) becomes available.

Exhibit 14. Building Detailed Billing Records from Multiple Sources

Vendor	Summary Number	Account Number	Bill Number	Bill Amount	LD Charges	Current Charges	Bill Date	Circuit Type_ID
Bell Atlantic	201 X52–0066–999 32	201 291–0001–810	201 291–0801	$21.79		$21.79	1-Jan-01	
Bell Atlantic	201 X52–0066–999 32	201 325–0002–073	201 325–8170	$41.72		$41.72	1-Jan-01	IBZD876332...NJ
Bell Atlantic	201 X52–0066–999 32	201 325–0003–073	201 325–8355	$0.00		$0.00	1-Jan-01	
Bell Atlantic	201 X52–0066–999 32	201 467–0004–167	201 467–3466	$21.79		$21.79	1-Jan-01	
Bell Atlantic	201 X52–0066–999 32	201 533–0005–033	201 533–0920	$46.79		$46.79	1-Jan-01	IBZD874983...NJ
Bell Atlantic	201 X52–0066–999 32	201 533–0006–033	201 533–0981	$0.00		$0.00	1-Jan-01	
Bell Atlantic	201 X52–0066–999 32	201 628–0007–075	201 628–4977	$46.14		$46.14	1-Jan-01	IBZD870434...NJ

Step 7: Initiate a Circuit Utilization Analysis

Objectives

Identify opportunities to eliminate excess capacity and consolidate or restructure traffic. This is the most detailed and technical portion of the project.

Process

A first-cut analysis uses a "what-if" approach based on common-sense network knowledge. A more formal analysis with optimization tools can be used to identify greater opportunities. Analysis steps include the following:

- Document performance requirements for each location and category of application. Use quantitative measures such as response time, reliability, and number of alternate paths. When necessary, incorporate qualitative information.
- Develop a straw-man set of requirements based on an understanding of the operation and experience at other sites when specific performance information is not available. Incorporate comments from internal staff and continue the process until all parties agree to the performance criteria.
- Obtain information about the utilization of voice circuits from the PBXs. Be aware, however, that different PBXs provide the data in varying formats. This information is useful as input to formal optimization analysis as well as eyeball analysis.
- Collect information regarding the utilization of data circuits using tools such as HP's TRENDsnmp+ package. TRENDsnmp+ collects the information from the routers (see Exhibit 15).
- Use the reports to understand the utilization on each circuit as a percentage of capacity. Review the three-day report to see average, maximum, and 90-percentile utilization in each direction. Use the one-month report to see average, minimum, and maximum utilization in each direction.
- Obtain information regarding usage and business need for circuits from other sources such as sales organizations.
- Determine which circuits are candidates for reconfiguration, using the performance requirements, utilization data, and business needs information collected.
- Prepare all utilization and application requirement data and load into the planning tool.
- Run the planning/optimization tool with as many options as are necessary to provide alternative circuit configurations.
- Incorporate the formal optimization information obtained to identify potential savings from restructuring the wide area network.

General Comments

- If usage reports include circuit ID, that information can be used to cross-reference utilization data with billing and inventory data. Unfortunately, the reports do not always include circuit ID.

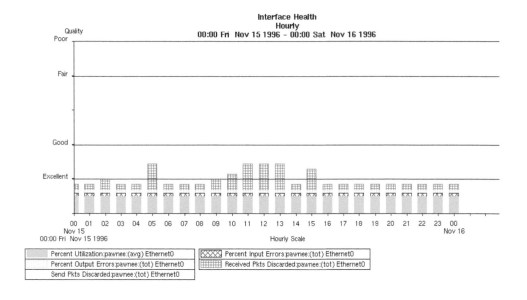

Exhibit 15.   Sample TRENDsnmp Report Showing Grade of Service and Potential Network Congestion
(Courtesy of Hewlett-Packard Company.)

- Network optimization studies often identify 15 to 20 percent savings but require complete documentation of the traffic, utilization, and performance requirements.
- Low usage does not mean the business need is minimal. Low-usage circuits may be serving as backup in case primary circuits fail, or direct high-speed circuits between two sites may satisfy response time and performance requirements even if volume remains low. Sometimes single line circuits are used for security alarms or elevator phones. Furthermore, cancellations should not be based solely on current utilization. Plans for expansion and the need for alternate paths should be considered.
- Circuits for a firm's small offices usually represent a relatively small percentage of the enterprise total network costs. Accordingly, analysis of these locations can usually be deferred without a major effect on total savings.
- Any delay in acquiring tariff and optimization tools inhibits the ability to reconcile billing inconsistencies and identify savings.

Sample Requirements Information for a WAN Analysis

Exhibit 16 shows an example of information needed for WAN optimization. The V and H (vertical and horizontal) coordinates shown in this exhibit are used by long-distance carriers in the United States to calculate the mileage between points. NPA-NXX are the first six digits of a North American telephone number, that is, the area code and exchange.

Sample Requirements for Point-to-Point Voice Network Analysis

Exhibit 17 is an example of a voice network (point-to-point) optimization worksheet. Converged voice-data networks will be addressed later.

Exhibit 16.   WAN Optimization Worksheet

Data	Comment	Example
Chicago		**Site 1**
NPA	Need NPA-NXX, or V and H, not both	201
NNN	Need NPA-NXX, or V and H, not both	265
H	Need NPA-NXX, or V and H, not both	
V	Need NPA-NXX, or V and H, not both	
Chattanooga		**Site 2**
NPA	Need NPA-NXX, or V and H, not both	201
NNN	Need NPA-NXX, or V and H, not both	291
H	Need NPA-NXX, or V and H, not both	
V	Need NPA-NXX, or V andH, not both	
(Traffic A to B)		
Peak hour	Time of day	9AM
Peak hour bits/sec	Includes raw data and overhead	532000
Peak hour packets/sec	Includes data and admin packets	623
Typical hour	Time of day	3PM
Typical hour bits/sec	Includes raw data and overhead	24560
Typical hour packets/sec	Includes data and admin packets	367
(Traffic B to A)		
Peak hour	Time of day	11AM
Peak hour bits/sec	Includes raw data and overhead	65480
Peak hour packets/sec	Includes data and admin packets	123
Typical hour	Time of day	3PM
Typical hour bits/sec	Includes raw data and overhead	8754
Typical hour packets/sec	Includes data and admin packets	12
Requirements		
Average delay A to B	Average delay/packet must be less than this value	0.5 sec
90th percentile delay A to B	90 percent of delays must be less than this value	1.5 sec
Utilization A to B	Average utilization in peak hour must be less than this value	0.6
Average delay B to A	Average delay/packet must be less than this value	0.5 sec
90th percentile Delay B to A	90 percent of delays must be less than this value	1.5 sec
Utilization B to A	Average utilization in peak hour must be less than this value	0.6
Diversity	Number of alternate routes between A and B must be ≥ this value	2
Max. hops	Number of hops between A and B must be ≤ this value	3
Reliability	Percent of time connection is up must be ≤ this value	0.9999

Exhibit 17. Voice Network Optimization Worksheet

Data	Comment	Example
Chicago		**Site 1**
NPA	Need NPA-NXX, or V and H, not both	201
NXX	Need NPA-NXX, or V and H, not both	265
H	Need NPA-NXX, or V and H, not both	
V	Need NPA-NXX, or V and H, not both	
Chattanooga		**Site 2**
NPA	Need NPA-NXX, or V and H, not both	201
NXX	Need NPA-NXX, or V and H, not both	291
H	Need NPA-NXX, or V and H, not both	
V	Need NPA-NXX, or V and H, not both	
(Traffic A to B)		
Peak hour	Time of day	9 a.m.
Peak hour usage, in Erlangs		23
Typical hour	Time of day	3 p.m.
Typical hour usage, in Erlangs[a]		7
(Traffic B to A)		
Peak hour	Time of day	4 p.m.
Peak hour usage, in Erlangs		17
Typical hour	Time of day	10 a.m.
Typical hour usage, in Erlangs		5
Requirements		
Blocking probability	Probability of no trunk avail must be ≤ this	0.01
Diversity	Number of alternate routes between A and B must be ≥ this	2
Reliability	Percent of time connection is up must be ≥ this	0.9999

[a] An Erlang equals one hour of voice communication. Erlangs indicate the quantity of voice traffic.

Sample Application Description

Exhibit 18 illustrates application-level information that should be collected. This information will vary based on the application and needs of the organization.

Sample Straw-Man Information

Exhibits 19 through 21 provide straw-man examples starting with location information, workload, and application data, respectively.

Exhibit 18.   WAN Application Worksheet

Data	Comment	Example
Application name		Customer information system
Application description		Get customer info
Application type	Examples: batch, interactive, LAN	Interlan
Protocol	Examples: SNA, TCP/IP, IPX	IPX
Protocol overhead	Either percent or constant per packet	50 percent
Transmission size	Indicate if data only	200 bytes
Response size	Or data and overhead	10,000 bytes
Response time requirement	Maximum time for communications response, not including processing time	5 sec
Backup requirement	Number of alternate paths required if the application consists of several different kinds of "transactions," provide info for each, and indicate relative frequency	0
For each source-destination pair that uses this application		
Source		
Destination		
No. "transactions"/peak hour	Used to get peak hour traffic	50
Peak hour	Busiest hour for this app	4 p.m. last day of month

Exhibit 19.   Location Descriptions

Node Name	NPA	NXX	Description	# of Users
Site 1	865	261	Headquarters	1
Site 2	615	712	Site 6, Site 7	1
Site 3	901	288	Site 8	50
Site 4	212	664	Site 9	50
Site 5	212	709	Sales office	10

Exhibit 20.   Workload Descriptions

Workload Source	Destination	Description	Trans/Hour	User Bits/Hr Average	Calculated User Bits/Sec	Router Total Bits/Sec
Site 1	Site 2	HQ to Site 7	20,000	160,000,000	44,444	90,090
Site 1	Site 3	HQ to switch	100,000	80,000,000	22,222	58,672
Site 1	Site 4	HQ to switch	100,000	80,000,000	22,222	94,184
Site 1	Site 5	HQ to Sales	10,000	8,000,000	2,222	49,408
Site 1	Site 10	HQ to switch	100,000	80,000,000	22,222	1,544
Site 1	Site 11	HQ to switch	200,000	160,000,000	44,444	50,952
Site 1	Site 1	HQ to Customer Care	40,000	1,280,000,000	355,556	810,810
Site 1	Site 12	Local support	20,000	160,000,000	44,444	117,560
Site 2	Site 1	Site 7 to HQ	20,000	480,000,000	133,333	270,270
Site 2	Site 6	Site 7 to switch	25,000	10,000,000	2,778	4,632

Exhibit 21.   Application Descriptions

Name	Burstiness (Max./Avg.)	User Data per Trans Bytes	Response Time Required (sec)	Protocol	Requires Alternate Path	Trans per "User" per Hour
Switch to Site 7	5	100	0.1	IP	Yes	500
Site 7 to Switch	10	50	0.1	IP	Yes	500
Switch to HQ	13	100	1	IP	Yes	2,000
HQ to Switch	14	100	1	IP	Yes	2,000
HQ to Site 7	10	1000	1	IP	Yes	20,000
Site 7 to HQ	2	3000	1	IP	Yes	20,000
Cust. Care to Site 7	2	2000	2	IP	Yes	10,000
Site 7 to Cust. Care	2	1000	2	IP	Yes	10,000
Sales to HQ	15	100	5	IPX	No	1,000
HQ to Sales	13	100	5	IPX	No	1,000
Cust. Care to HQ	4	4000	1	IP and IPX	Yes	40,000
HQ to Customer Care	4	4000	1	IP and IPX	Yes	40,000
local support (C to S)	20	500	1	IPX	No	1,000
Local support (S to C)	15	1000	1	IPX	No	1,000

Note: The following are believed to have no traffic:

    Switch to Customer Care
    Customer Care to Switch
    NOC to Lab
    Lab to NOC
    Customer Care to Lab
    Lab to Customer Care

Exhibit 22.   Telecom Activities Reported by Intelicontrol Package from Switchview

Multi-level traffic analysis
Least-cost routing analysis
Comprehensive alarm management
Coordinated dialing plan
Work-order processing
Automatic moves, adds, and changes
Networkwide call accounting
Integrated directory
Full inventory management
Equipment and service cost
Allocation
Cable tracking
Automatic directory updating
Online listing delivery
Attendant call management
Desktop dialing
Toll-fraud detect and disable
Scheduled trunk shutdown
Auth-code monitoring

Sample Switch View Report

In recent years, telecom management packages have proliferated and now offer a wide range of options. For example, the Intelicontrol package from Switchview (www.switchview.com) reports on telecom activities such as those listed in Exhibit 22.

Exhibit 23 illustrates network optimization data that is available from Switchview.

All the major PBX vendors supply trunking information directly from console commands. Exhibit 24 illustrates some of the detailed capacity planning and utilization information from Avaya's Definity PBX.

Sample Telco Circuit Utilization Request

In cases where internal tools are not available or the current tools cannot obtain traffic utilization data for certain circuits, the telcos may be able to supply the missing circuit information. The letter shown in Exhibit 25 illustrates information that would typically be requested for a network analysis.

Step 8: Realize Savings

Objectives

Identifying savings (at last!). Following is the process to identify, categorize, and realize network cost savings:

Exhibit 23. Sample Network Optimization Data from Switchview

Trunk Route Call Details (SMDR)

Trunk Route Number: 003
Trunk Route Type: CO
Facility: Site NN/15E
Grade of Service (G.O.S.) Model: Erlang B
Target Grade of Service: 1.00 percent

Hr. Toll Calls	Call Peg Counts			Usage			Average Time	Holding Util.	Percent G.O.S.	Actual Trunks Eqp'd	Trunks Not Used	Required	Trunks
	Incoming	Outgoing	Total	Incoming	Outgoing	Total							
Study Date: 04/14/99													
Day of Week: Monday													
9	0	2	2	:00	:11	:11	:11	5:50	0.85	0.00%	23	22	3
10	0	2	2	:00	:00	:00	:00	0:00	0.00	0.00%	23	22	0
11	2	4	6	:20	:03	:23	:23	3:53	1.69	0.00%	23	19	3
...													
Study Date: 04/15/99													
Day of Week: Tuesday													
9	0	2	2	:00	:11	:11	:11	5:50	0.85	0.00%	23	22	3
10	0	2	2	:00	:00	:00	:00	0:00	0.00	0.00%	23	22	0
11	2	4	6	:20	:03	:23	:23	3:53	1.69	0.00%	23	19	3

Exhibit 24. Sample Trunking and Utilization Information Available from Avaya's Definity PBX

Measurements trunk-group summary yes: 04/26/97:   45   Page 1   SFE A

Switch Name: Date: 6:45 am SAT APR 26. 1997

Trunk Group Summary Report

Peak Hour for ALL Trunk Groups: 1500

Grp No.	Grp Siz	Grp Type	Grp Dir	Meas Hour	Total Usage	Total Seize	Inc. Seize	Grp Ovfl	Que Siz	Call Qued	Que Ovf	Que Aod	Que Srv	ATB	Que ST
1	94	isdn	two	1000	662	352	352	0	0	0	0	0	0	0	0
2	96	tie	two	2100	990	429	1	0	0	0	0	0	0	0	0
3	116	isdn	two	1500	2134	641	641	0	0	0	0	0	24	0	0

Command successfully completed

Process

- Conduct a brainstorming session with team members to create a list of potential circuit savings categories.
- Develop the criteria that would place an individual circuit into the appropriate category.
- Create lists of key circuit fields by each category based on defined criteria.
- Evaluate active circuits and obtain billing information (e.g., summary account, circuit ID, latest monthly charges, and bill group).
- Factor in the previous circuit categories and priority requirements.
- Give consideration to backup requirements
- Identify circuits that are billed but not connected.
- Review the final circuit list for cancellation, moves, etc. with the appropriate responsible person(s).
- Obtain authorization sign-off.
- Notify the telco.
- Track the service order and telco response on the disconnect date. See Exhibit 26 for an example of a tracking form used to follow up on the telco status.
- Follow up termination of billing service.

General

- Continually follow up with the telcos to ensure that cancellations are processed and completed. Also keep track of credits due.
- Dedicate staff to follow-up and tracking efforts. Continual vigilance is required.
- The following are examples of potential savings:
  - Closed sites: company sites that have physically closed but still have circuits to other sites
  - Zero usage or closed retail sites: usually circuits that are attached to retail sites that no longer exist or have no usage over some period of time
  - Work at home: circuits that have not yet been disconnected to a former employee's home
  - Zero or low usage: circuits that are not to retail sites but between other site facilities; some of these circuits are no longer required, causing an imbalance in capacity
  - Restructure voice or data circuits: restructuring of site circuits based on criteria such as application requirements, backup requirements, redundancy, and usage
  - Renegotiate contracts: circuits supplied by a vendor in a long-term contract that go to sites which are either closed or have no traffic on them; the penalties involved through early cancellation are generally small, providing the opportunity for some significant savings
  - Out of region: circuits charged to one region that should be charged to another; any reassignment of circuits, as well as a credit, provides budgetary savings

Exhibit 25.   Sample Information Request to a Carrier for Circuit Information

March 11, 2002

Mr. John Doe
XYZ Carrier
Corporate Account Manager

Customer Service Department

Dear Mr. Doe:

ABC Widget Company is conducting an analysis of its circuits provided by your company. Please provide the following information regarding our circuits:
- For voice circuits:
  - Circuit or group identifier, as identified on order records/billing records sent to firm
  - Number of circuits in group
  - Source location: address and NPA-NXX
  - Destination location: address and NPA-NXX
  - Measurement dates and times (start and end): see instructions below
  - Average utilization, in Erlangs
  - Average blocking probability
  - Number of hours of circuit outage
  - Peak hour (date and time)
  - Utilization for peak hour
- For physical data circuits:
  - Circuit identifier, as identified on order records and billing records sent to firm
  - Data rate, in bits per second (bps)
  - Source location: address and NPA-NXX
  - Destination location: address and NPA-NXX
  - Measurement dates and times (start and end): see instructions below
  - Average utilization for source to destination traffic
  - Average utilization for destination to source traffic
  - Number of hours of circuit outage
  - Peak hour (date and time)
  - Utilization for peak hour, source to destination
  - Utilization for peak hour, destination to source
- For logical data circuits, such as Frame Relay PVCs:
  - Circuit identifier, as identified on order records and billing records sent to firm
  - CIR, in bits per second (bps)
  - Maximum burst rate, in bits per second (bps)
  - Source location: address and NPA-NXX
  - Destination location: address and NPA-NXX
  - Measurement dates and times (start and end): see instructions below
  - Average utilization for source to destination traffic
  - Average utilization for destination to source traffic
  - Number of hours of circuit outage
  - Amount of burst rate traffic, source to destination
  - Amount of burst rate traffic, destination to source

**Exhibit 25.   Sample Information Request to a Carrier
for Circuit Information (Continued)**

- Peak hour (date and time)
- Utilization for peak hour, source to destination
- Utilization for peak hour, destination to source
- Amount of burst rate traffic, source to destination
- Amount of burst rate traffic, destination to source

The measurement dates and times should span at least one month, although a three-month period is considerably more useful.

Please contact me at (713) 123–1234 if you have any questions. Thank you for your attention to this matter.

Sincerely,

William Alcabiades
ABC Widget Company Telecom Analyst

- Application retirement: any application that had circuits and equipment that are no longer used (replaced by another application or different types of circuit services)

See Appendix A for a more complete list of potential sources of savings.

Exhibit 26 shows an example of a tracking form for a closed site, used to ensure that the telco completes assigned shutdown activities.

Step 9: Reconcile and "Baseline" Circuit Inventory

Objectives

Use the consolidated information (from Step 6) and reconcile the information by comparing billing information to network inventory data. After reconciliation, a baseline circuit inventory is developed that is as accurate and as up-to-date as possible.

Process

- Decide which data elements (fields) to compare between billing and circuit inventory information. For example, circuit or BTN (bill to number) could be compared.
- Create a single reconciliation table that contains all the fields from the billing and inventory tables as the target table for matches.

Exhibit 26. Telco Follow-Up Form

Account Number	AT&T	Bell Atlantic	Service Order Date	Disconnect Date	Monthly Savings	Site/Loc	Comments
8000–74110–26	DHEC301106	HCGS066695	4/30/01		$715.54	111 W.	LP2 BG = 734
8000–74110–26	DHEC301107	HCGS066696	4/30/01		$715.54	111 W.	LP9 BG = 734
8000–74110–26	DHEC301108	HCGS066697	4/30/01		$715.54	111 W.	LP10 BG = 734
8000–74110–26	DHEC301109	HCGS066698	4/30/01		$715.54	111 W.	LP18 BG = 734
8000–74110–26	DHEC226608.100	HCGS066730	4/30/01		$715.54	111 W.	LP19 BG = 734
8000–74110–26	800–841–8821		4/30/01		$257.08	111 W.	And all assc. meas. ports (46) BG734
8000–74110–26	800–841–8853		4/30/01		$257.08	111 W.	And all assc. meas. ports (48) BG734
201-V08–0516		DHSA855408	4/23/01	4/28/01	$620.00	345 N..	
201-V08–0516		DHSA855409	4/23/01	4/28/01	$620.00	345 N.	
201-V08–0516		DHSA855410	4/23/01	4/28/01	$620.00	345 N.	

Exhibit 27. Consolidated Table of Billing Records

Billing Vendor:	AT&T	
Account Number:	8000–54509–94	
Billing Number:	415270	FO 741
Circuit Number:	AQEC922253	ATI
Total Amount:	$618.81	
Bill_"Firm"_Category:	Org- BILL GRP 741	
Bill_Application:	X & Y Electronics	

- Using consolidated tables as well as billing and circuit tables, compare records based on the developed criteria.
- Change the above match query to a not-in logic on a small billing table comparison to the larger circuit inventory table. This results in a billing table of noninventory matches.
- Update the unreconciled billing table(s) by obtaining circuit-related information from such sources as service order forms, CD-ROM circuit inventory data, spreadsheets, and telco CSRs. Based on the types of circuit information obtained, add these fields to the billing table. The updates may require both manual entry and reformatting of electronic source data. Missing circuit information may hamper reconciliation of some of the records.
- Copy the updated billing table records into the single reconciliation table mapping the appropriate billing table fields into the reconciliation table.
- Create a history table containing the same fields as the reconciliation table.
- Delete from the reconciliation table any circuit disconnects that result from the savings realization effort. Copy these records to the history table.
- Modify the reconciliation table with all new circuits added or circuits with change service orders. On the change circuits, save the original circuit record into the history table.
- Reflect other circuit information brought to light, such as the discovery of more current bill details in the reconciliation tables.

General Comments

- Ensure that appropriate backup tables are created for most of the identified process steps.
- Identify the point of diminishing returns for reconciliation and baseline development.

Sample of a Reconciled Event

Exhibit 27 provides the type of information contained in a consolidated table of billing records. From the information above, match the circuit inventory table to the circuit number field. If possible, avoid matching against account

**Exhibit 28. Additional Circuit Information
from the Network Circuit Inventory Table**

Ntwk_C_OR_ID_NO	AQEC922253
Ntwk_APP_DESC	Org 9.6 LINE
Ntwk_SITE_NAME	X&Y ELECTRONICS
Ntwk_STREET_ADD	1500 NNN DRIVE
Ntwk_CITY	"WESTBURY,"
Ntwk_STATE	NY
Ntwk_ZIP_CODE	
Ntwk_REMOTE_CON	SAM GOTTLEIB
Ntwk_ATT_W_CON	ERIC GEARD
Ntwk_ORIGINATES	"WESTBURY, NY"
Ntwk_TERMINATES	"EDISON, NJ"
Ntwk_DIRECTION	INBOUND TO HQ
Ntwk_LINE_TYPE	ANALOG
Ntwk_SPEED	9.6
Ntwk_INSTALDATE	5/26/99
Ntwk_NOTES	X&Y 9.6 CIRCUIT FOR ORG.
Ntwk_BILL_NUM1	415270 FO 741
Ntwk_CKT_ID1	AQEC922253
Ntwk_CKT_USE1	DATA
Ntwk_ADDR1	FLR 2
Ntwk_ADDR2	DATA CENTER
Ntwk_ADDR3	NNN EAST ABC AVE
Ntwk_ADDR4	EDISON NJ ZIP
Ntwk_NPA1	201
Ntwk_NNN1	999
Ntwk_ACATLCO1	RRRRNJ02
Ntwk_SWC_CODE1	SSSSCA01
Ntwk_BILL_NUM2	415270 FO 741
Ntwk_CKT_ID2	AQEC922253 ATI
Ntwk_ADDR1B	FLR MAIN
Ntwk_ADDR2B	DATA CNTR
Ntwk_ADDR3B	1500 NNN DR
Ntwk_ADDR4B	WESTBURY NY11590
Ntwk_NPA2	516
Ntwk_NNN2	999
Ntwk_ACATLCO2	GRCYNYGC
Ntwk_SWC_CODE2	SSSSCA01

number or bill number because they are not unique, and such matches would result in false information that must be manually removed. When matching the circuit number field with the ntwk_c_or_id_no field from the network circuit inventory table, additional circuit information can be obtained, as shown in Exhibit 28.

Exhibit 29.   Billing Record with No Match in the Network Inventory Table

Bill_Vendor	AT&T
Account Number	8000–75509–94
Bill Number	415270 FO 741
Circuit Number	DHEC555120100ATI
Total Amount	$0

Exhibit 30.   Field Information Available from AT&T Billing CD-ROM

Ntwk_C_OR_ID_NO	DHEC555120100ATI
Ntwk_VENDOR_LD	ATT
Ntwk_BILL_NUM1	415270 FO 741
Ntwk_CKT_ID1	DHEC555120100ATI
Ntwk_NPA1	201
Ntwk_NNN1	999
Ntwk_ACATLCO1	RRRRNJ02
Ntwk_BILL_NUM2	415270 FO 741
Ntwk_CKT_ID2	DHEC555120100ATI

Sample of an Unreconciled Event

Exhibit 29 shows information contained in a billing record with no match in the network inventory table. In this case, circuit information must be obtained by other means. Using an available AT&T Billing CD-ROM, the field information shown in Exhibit 30 is available.

Step 10: Assess Processes

Objectives

Identify cost savings and improve the accuracy of telecom data.

Process

- Interview individuals to determine their perspectives on gaps or shortcomings related to network administration. Include those responsible for inventory in a particular segment of the network, accounts payable, requesters within or outside the organization, the order administrator, network analysts, and telco contacts.
- Hold a team meeting to discuss the findings, obtain consensus, and develop possible solutions.

- Jointly review and document the organization's order fulfillment process.
- Evaluate the types of information captured versus what is necessary to improve areas such as network knowledge, maintenance, trouble tracking, and control.
- Evaluate the organization's infrastructure to form teams, assign responsibilities, and identify critical gaps.

General Comments

- Create a part-time network oversight committee to coordinate, focus, and overall manage network growth. Areas to address include the creation and introduction of process standards, change management, network strategies, and business needs alignment.
- Create a team of senior network representatives from both the voice and data organizations to evaluate all new, change, and delete orders before they are placed with the telcos. This team is responsible for the recommended action items below:
  - Conduct biweekly meetings to review all new orders before any telco request is made.
  - Review the request for authorization and project application information.
  - Evaluate the request to determine the effect on the project and approve or disapprove.
  - Look at the potential to reengineer the request. Evaluate the request against any build or application plans. When possible, incorporate the request into those plans.
  - Approve or disapprove any petition to expedite the request.
  - Identify any required equipment additions or modifications.
  - Create any additional telco specifications.
- When assessing an order fulfillment process, consider the following areas:
  - Volume and time requirements for processing service orders
  - Information is needed on the request form. How will the organization communicate with the supplier for service order requests, provisioning, and confirmation?
  - Network engineering input to the service order process.
  - The segregation of operational requests in the service order process.
  - A tool that captures requester data, telco information, and circuit information (create it if necessary). The tool should feed a database table as well as create circuit request forms, a database for capturing order, provisioning, and billing information.
  - Bill reconciliation and chargeback mechanisms.
  - Business requirements.
  - Network planning.
  - The network information management system.
  - Detailed methods and procedures for each functional step in the order fulfillment process. The new process should be advertised and adhered to by all affected organizations (e.g., billing, design team, etc.).

Exhibit 31 illustrates a process redesign for current and new order fulfillment.

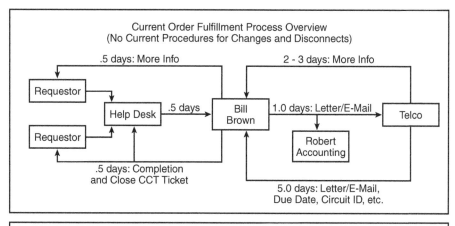

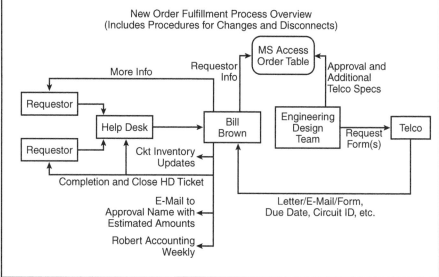

Exhibit 31. Sample Order Fulfillment Process Redesign

Step 11: Maintain the Gains

A network inventory and optimization analysis project often uncovers significant opportunities for improving processes, increasing operational controls, and realizing savings. To avoid repeating the inventory reconciliation as a frequent clean-up effort, an organization should take steps to maintain the gains made during the project.

There are three major components to ongoing management of network expenditures and control of inventory:

1. *Billing and circuit inventory reconciliation.* Billing and circuit inventory records must be consistently monitored to reflect what is actually operational and necessary to meet the needs of the firm.

2. *Network planning and optimization.* Network planning (e.g., voice and data, local and long distance) is required to take advantage of economies of scale and other synergies. Avoid ad hoc changes to the network in response to localized short-term needs.
3. *Process enhancement follow-up.* Process enhancement recommendations require follow-up. Moreover, it is critical that staff commitment is solidified and focused on the alignment of business needs, ongoing analysis and tracking, and the implementation of new tools.

Billing and Circuit Inventory Reconciliation

An accurate inventory, so painfully accumulated, will tarnish like buried treasure if not maintained. Following is a list of good practices and observations that will help maintain the value:

- A pricing tool, based on supplier tariffs, should be used to help identify billing errors.
- Reconciliation of the billing and inventory should be maintained because so much effort went into the process.
- Circuit data inconsistencies should be identified and continually monitored to help lower excess costs.
- Billing and inventory databases and processes should remain synchronized. While consistency requires considerable effort, the risk of divergence is additional cost, unnecessary effort, and the loss of valuable gains recently achieved.
- Frequent oversight review should ensure that the firm's telecommunication needs are reflected in the network. For example:
  - When a site is closed, follow-up is necessary to ensure that all the relevant circuits are canceled and that the associated long-distance charges are terminated.
  - When an application changes, a review of circuits is required to ensure those circuits are still needed.
  - When an employee leaves the company, a linkage to the Human Resources department is required so that any work-at-home circuits are canceled.
- Because increasing day-to-day operations (the "urgent") distract telecommunications and financial staff, more senior individuals should periodically conduct reviews.

The implementation and oversight of these activities may require the addition of one FTE (full time equivalent), depending on organization size.

Network Planning and Optimization

Project gains depend on network planning and management to achieve optimization:

■ Thorough network planning encourages the ability to realize considerable savings, particularly when billing and inventory remain synchronized and incorporate the firm's business requirements. The following recommendations may provide additional potential savings:
  - Consolidate multiple low speed circuits onto more cost effective high speed circuits
  - Consolidate voice and data onto a single high-speed circuit.
  - Reconfigure data circuits so that traffic can share existing circuits and avoid dedicated connections, where utilization is low, and applications can tolerate the minimal additional delay. For example, reconfigure traffic so that instead of going from A to C, have it go through B rather than having its own dedicated connection.

■ Formal planning tools using optimization algorithms and a tariff database can identify cost savings while allowing the ability to achieve network throughput, response time, grade of service, and redundancy requirements.

■ Performance requirements must be documented and agreed to by all concerned parties for both formal and informal (e.g., "eyeball") planning.

■ In the absence of documented requirements, a straw man can be developed based on experience and knowledge of the operations.

Process Enhancements

The following points summarize key steps in maintaining cost-effective WAN communications:

■ Create a part-time network oversight committee to coordinate, focus, and manage growth. Specific areas to address include the creation and introduction of process standards, the implementation of change management, the incorporation of network strategies, and the alignment of business needs.

■ Create detailed methods and procedures and communicate them to all organizations supported by the firm's network group.

■ Decide early how to make best use of captured data. This information may require a migration or conversion to an improved operational network management inventory system.

■ Coordinate with the supplier to develop a streamlined process for placing orders, provisioning, and obtaining confirmation content.

■ Investigate chargeback mechanisms on administrative network components (e.g., implement CD-ROM or EDI for billing). This allows proper usage, authorization, and cost allocation between the firm's organizations.

■ Implement an improved circuit database. In concert with any process improvements, an improved circuit database should be implemented that can be easily maintained, provides meaningful management reports, and incorporates interfaces to other systems such as ordering, provisioning, and trouble management, planning tools, and billing.

■ Continually investigate, track, and monitor the savings realization activity for the previously identified savings categories.

Summary

The combination of technical analysis, review of existing information, and process improvements can almost certainly reduce wide area networking expenses for most organizations. While there is certainly a significant effort involved, the payoff continues to accrue over time.

Notes

1. X.25 is a much slower data communications standard that provides for massive error checking between transmission points. It is still used in many parts of the world. Frame Relay, in contrast, performs minimal error checking (it is "unreliable" in itself). However, it achieves the same result by relying on the error correction capabilities of edge equipment and higher layer protocols; hence its much greater speed than X.25.
2. Frame Relay Access Device. A FRAD handles header and trailer frames. It is sometimes a stand-alone device but more often is included in the functionality of a router.
3. MPLS (multi-protocol label switching) provides Internet service providers with the ability to reroute packets around failed circuits, congestion, and bottlenecks.
4. George Gilder, From Wires to Waves, *Forbes ASAP,* June 5, 1995.
5. These are brief disruptions of satellite reception caused by an exact alignment of the sun and satellite relative to the VSAT receiving dish. These disruptions only occur during the autumnal and vernal equinoxes.
6. Cellular digital packet data (CDPD). Supports transmission speed up to 19.2 kBps over the older analog cellular network. CDPD takes advantage of pauses in voice communications and transmits packets during that time. The speed often drops well below 19.2 during peak traffic.
7. Structured query language (SQL). SQL statements are used to perform database tasks such as update data or retrieve data based on specified criteria.

Chapter 5

Architecture and Local Access Strategies

Knowledge is of two kinds. We know a subject ourselves, or we know where we can find information upon it.

— James Boswell, *Life of Samuel Johnson*

The architecture of an organization's voice communications system directly affects the total cost of operations. Options vary from the obvious — whether to use voice-over-IP (VoIP) — to the subtle, such as appropriate options for least cost routing. It is generally recognized that most networks will evolve into one combined voice and data architecture. Recently, the TeleGeography 2002 report (www.telegeography.com) showed a six percent VoIP penetration worldwide by the end of 2001 (ten billion minutes of business VoIP traffic). Meanwhile, choices must be made and money spent. The sections below present options for an enterprisewide PBX (voice communications) architecture, local access alternatives, and related costs.

Voice-over-IP: Premises Equipment

Voice-over-IP (VoIP) has two components: (1) premises equipment, such as IP telephones and switches, and (2) wide area network facilities and equipment, such as gateways and VoIP cards in routers. Popular publications sometimes confuse the subject by not distinguishing between the two. In some cases, an IP-based PBX is cost-effective when VoIP over a wide area network is not, and vice versa.

A premises-based IP PBX[1] has the following advantages over the traditional, proprietary PBX:

- There is the potential to use a single wiring network for both data and voice rather than separate, parallel wiring systems now used.
- Ease of move, add, change (MAC) within a building. Instead of extensive administrative changes via a proprietary interface, the user's telephone is simply moved from one office to another. Because each telephone set has its own IP address, it rings at the right place as long as it is on the organization's IP network. Some organizations spend thousands of dollars a month on moves within a single building.
- Newer applications, such as unified messaging,[2] are easier to implement on IP-based systems.
- Capacity can be added in much smaller increments than with traditional PBXs, which may require a new shelf, node, or even a complete forklift upgrade.
- IP-based PBXs are generally more open and standards based, holding out the promise of less costly hardware and software enhancements.
- Web-based applications can be easily linked with the IP telephony world. For example, it is far less expensive to install "screen pops" in an IP telephony environment than in traditional PBX systems. Sales personnel and others who need information on the caller can benefit from such features. Another example: an employee has a question about a 401K feature. He finds the information in the organization's intranet Web page. He clicks on a "click to talk" button and the appropriate party is dialed as he picks up his telephone to talk.

Despite these advantages, the IP PBX has a few limitations (at least for the moment):

- IP PBXs have not yet scaled to thousands of users. Nonetheless, each year the number of ports available on a single unit (such as the 3COM NBX 100) continues to increase.
- The "tank-like" reliability of traditional vendors such as Avaya, Nortel, and Siemens has not been achieved — at least in public perception. Like PCs in the early days of the mainframe world, IP PBXs have to evolve bullet-proof armor before Fortune 500 firms will trust their corporate headquarters' voice system to a new technology.
- The huge installed base of legacy, proprietary PBX software will need to be ported or developed for the IP world. That process is occurring rapidly.
- Organizations with unreliable wiring or whose current bandwidth is nearly occluded with data communications traffic (e.g., large file transfers) may not have the building infrastructure to support voice over data.

Exhibit 1 displays a simplified diagram of an IP-based PBX. Note that multiple links are enabled — traditional circuit switched (TDM) telephony, IP telephony via the LAN, and links to the Internet.

Exhibit 1. Example IP (LAN)-Based Telephone System
(Courtesy of AltiGen Communications.)

Although it is difficult to quantify the net economic effect, some of the features provided by LAN telephony contribute to greater employee productivity. Many of these features are available in traditional TDM telephony systems, but at a higher cost. Examples include:

- *Easy screen pops.* When a customer calls, a link is established between the incoming caller ID and a contact package such as ACT!, Outlook, or Goldmine. Information from the contact database is displayed immediately as the call comes in.
- *Call handling.* When an employee is on the line and another call comes in, a graphical interface simplifies decision making: the call can be ignored, accepted, routed to a queue for others to handle, sent to voicemail, or added to a conference. The key difference from past systems is that these features not only exist on less expensive platforms but they are often displayed on a workstation screen. Hence, employees can actually *use* the features on the system because the interface is simpler.
- *Simplified voicemail/unified messaging.* Users can use their workstation interface to listen, save, skip, delete, and scroll through voicemail messages. Clicking on a stored message can return calls. For those so inclined, messages can be saved as an Internet standard WAV file or more compressed proprietary file and forwarded as an e-mail attachment.

To further illustrate some of the capabilities and benefits of IP-based telephony, we can use the Siemens' optiPoint 100 advance IP telephone (ww.siemens.com) as a representative model. Siemens states that optiPoint provides for the following features and benefits:

- Features:
  - Hands-free and speakerphone
  - Memory dial and redial

- Display of the incoming number (CLI)
- Call hold/consultation
- Alternate
- Call forwarding (CFU, CFB, CFNR)
- Call waiting
- Call transfer
- Call deflection (user-controlled forward)
- CTI interface allowing TAPI client control
- Programmable ring tone, volume, and cadence
- Country-specific menu guidance

■ Benefits:
- Long-distance and toll calls can be transmitted over the IP network, reducing communication costs.
- Integrating voice and data into one network means investment in one technology and one support organization, reducing infrastructure costs.
- Software updates and feature enhancements can be downloaded quickly and easily, thus enabling cost-effective upgrades.
- Intuitive, interactive menu keys and displays along with simple dialing capabilities save time.
- Direct-dial keys are programmable, providing ease of use.
- OptiPoint 100 advance telephone automatically stores the numbers of the last 20 unanswered calls.
- Excellent voice quality in both hands-free and open listening modes using special digital signal processor (DSP) technology and acoustic algorithms for echo cancellation.

The Siemens IP phones can be upgraded with software (a good feature in a highly volatile technical environment). Other examples of IP telephones include InterPhone by DSG Technology and Cisco's IP phone 7960 (see Exhibit 2). Cisco's IP phone can also accept firmware updates via download.

When evaluating IP telephony solutions, voice quality is obviously critical. A standard measure, MOS (mean opinion score), is used by the industry to determine the subjective quality of the telephone conversation. The ranking is from 1 (very bad) to 5 (perfect, undistorted toll quality sound). One firm, NetIQ, has a well-developed monitoring system, VoIP Assessor, that allows simulation of VoIP traffic and an assessment of its quality. According to a recent *Business Communications Review* article, "The Assessor software measures delay, packet loss and jitter, and produces a report showing call quality by day of week, location, network cause, etc. ...There's been years and years of research that went into that ITU standard, so it really is a fairly scientific answer: You run this kind of traffic through this network with the parameters you told us, and here's what call quality's going to sound like." Call quality is expressed as MOS.

One clear indicator of the direction of the industry is the fact that Sprint, a major long-distance carrier, has decided to build out all its local telephone service using VoIP technology. Certainly the older technologies will co-exist for years, but the world is moving quickly to a fabric of interlacing packets that will carry information without regard to its original form.

Exhibit 2. Cisco IP Phone 7960
(Courtesy of Cisco Systems.)

The Centrex Alternative

According to *Newton's Telecom Dictionary,* Centrex is "basically normal single line telephone service with 'bells and whistles' added." While this is undoubtedly true for a large percentage of installations that want dial tone and perhaps a few extra services, Centrex is now offering more flexibility and features than in the past. Let us start with the basics. Why would an organization want to lease its entire infrastructure from the local telephone company rather than buy its own PBX? The most important reasons include:

- *Low start-up costs.* The equipment and software is owned, stored, and kept up-to-date by the carrier. Fees are a direct multiple of the number of stations and therefore highly predictable from year to year.
- *Ease of physical movement.* If employees move to new offices, the "virtual" telephone system easily moves with them. The risk of moves is diminished.
- *Cafeteria-style features.* Features are available on an ad hoc basis; users pay only for those they use.
- *High uptime.* Centrex service runs off central office-class PBXs (or softswitches) and therefore is unlikely to fail. Of course, the risk from cable cuts or other external factors is unchanged.

There are some disadvantages:

- *Higher costs for volatile environments.* Costs could easily escalate beyond an in-house PBX if users frequently move offices, require special services, or require special technology (e.g., computer telephony integration applications, such as screen pops).

- *Potential loss of service flexibility.* If the culture of the organization is "do it now," there could potentially be a conflict with operating under the provider's set schedule for moves, additions, etc. This disadvantage is somewhat lessened if the service provider offers IP telephones that can be logically moved by software.
- *May not scale.* While Centrex can certainly scale physically, it may not scale economically. As the number of users goes into the hundreds and thousands, the law of large numbers comes into play. Trunk lines can be reduced on a per-person basis, technologies and expertise can be spread over many more phones, and generally per-unit cost can be driven down.
- *Feature availability.* While Centrex offerings usually have the standard telephony features such as call forwarding, speed dial, park, etc., certain special features may not be available. If the organization wants something specific that is not a product offered, bringing telephony in-house may be required.

Small Offices and SOHO Markets: Key Systems versus PBX

For many small offices, a PBX is unnecessarily large and complex. Key systems will serve offices with less than 60 telephones; but beyond those numbers, a PBX is required. From a financial perspective, the "key" decision (pun unintended) is whether an office or plant will grow significantly, thus requiring a forklift upgrade to a PBX. Key systems cannot be easily scaled, although the distinctions between PBXs and key systems are less distinct now than in the past. Following are the traditional points of separation between key systems and PBXs:

- Each key system telephone has buttons or keys that are used to *directly* access external lines ("Harry, pick up line 24"). In a PBX environment, users are not linked to specific external lines, although they may have a second or third "line" on their telephone set. A key system may have 32 telephones and a dozen external phone lines.
- PBXs *share* trunking lines among a larger number of users. Using a DID (direct inward dial) number, a call is routed from an incoming trunk to the user's telephone. The PBX has the ability to switch rather than depend on a one-to-one user:line connection. This allows a much higher ratio of users to trunks and, with appropriate equipment, permits scaling into the tens of thousands of stations.[3]
- Although key systems continue to offer more features, PBXs still provide considerably more functionality, both from an end user and internal perspective.
- On a per-station basis, key systems can cost as little as $200 to $250; that number easily doubles for PBXs.
- The ability to select trunks can reduce costs in some areas of the country where LECs charge higher rates for pooled access trunks (i.e., those used by PBXs) than for lines that are selected by the user.
- PBXs have a robust ARS (automatic route selection) capability. By looking at part or all of the numbers dialed, the PBX can select the most cost-efficient

method of completing the call. For example, assume that a company has a headquarters building in Chicago and a smaller office in Denver. A user in Chicago dials the full ten digits to reach a co-worker in Denver. By looking at all ten digits, the PBX determines that the call should be routed via tie lines (or the IP equivalent if VoIP is in place) rather than the public network. It is important to have a PBX that can look at all ten digits because just looking at the area code and exchange may not be sufficient. For example, the Denver office might only have part of the exchange — the rest belongs to other organizations in the same area code.

- PBXs generally support full networking, including voicemail and a uniform dial plan (a uniform dial plan allows, for example, five-digit dialing across the United States and even worldwide; employees can reach each other with a minimum of dialing).

Key systems increasingly offer robust features that make them more attractive to the smaller office. In addition to their low cost, they offer functions such as:

- Voicemail
- Hunt groups. Incoming calls can be routed to a workgroup. Individuals will receive the calls in a predefined order and the call will be sent to the next available person in the sequence. In these cases, employee 1 on the hunt group list works hardest; employee 5 works least. Some key systems use smarter algorithms to distribute calls more evenly.
- Auto-attendant. Callers are sent to a main receptionist who then distributes the call as required. Computer-based auto-attendants provide the ubiquitous "Press 1 for Sally, 2 for Fred, etc."

The high end of the key system range is now the target of IP telephony vendors. However, the low end, say three to ten stations, continues to be dominated by traditional key system vendors. Because the low-end IP PBXs still require a reasonably robust server, there is a fixed cost that is difficult to overcome at the small end of the spectrum.

An example of the inexpensive (but not low-quality) key system is the Cortelco Aries 308 (see Exhibit 3). It accepts three CO (central office) lines and serves eight digital stations. Options for paging, external music on hold, and remote programming of features are available. If a small office needs only basic services, no inter-office system linking, and there are no rapid expansion plans, such a system may be appropriate.

Some key systems include broadband connections. For example, BizFon's 680 KSU (key system unit) has a 16-channel DSL card. Others support standard BRI (basic rate interface) and PRI (primary rate interface) links that allow more channels and potentially a discounted long-distance rate. The discounted rate is due to the ability of PRI trunks to directly connect to the long distance carrier, bypassing the local telephone company (this is sometimes termed "dedicated" access).

Exhibit 3.   Cortelco Aries 308 Key System

The Dreadnoughts: Enterprise Class PBXs

As of this writing, the majority of organizations with thousands of employees at one location continue to use traditional, TDM[4] enterprise class PBXs. For the next few years, these mainframe-like machines will continue to dominate the seas of telephony. They have extremely high reliability, more features than any person with a real job would ever use, and a vast flotilla of accessories/applications. The large installations cost millions.

Business needs should dominate technology "coolness." If the organization must have a new telephone system with all the advantages of mature technology, then cost analysis shifts away from technical architecture and toward features, needs, and smart planning. In evaluating competing vendors such as Avaya, Siemens, Nortel, and others that play in the large PBX world, the following should be included in the analysis:

■ Has an accurate count of users, stations, and non-voice devices such as modems and faxes been obtained? Organizations often overbuy because the actual number of necessary lines is not known. Analog lines (for faxes and modems) are particularly prone to excess.

■ Do voicemail and other proprietary links match the organization's infrastructure? For example, will the PBX vendor's unified messaging offering work with the installed e-mail system?

■ What functions are really required for each station? Some organizations buy low-end telephones for lower-level employees and then upgrade as they are promoted. Sometimes, the cost of this transition can exceed the cost of providing a standard, strong feature set telephone for all employees, with only individuals in specialty positions having "more buttons."

- How much expansion room is available? Traditional PBXs scale well but usually in steps rather than on a continuous curve of users versus costs. Once a shelf is filled up, for example, a new shelf and cards must be purchased, at a substantial incremental price. Of course, most of the big PBXs network with one another easily (as long as they are from the same manufacturer).
- How much of the implementation effort is the vendor willing to do? Implementation is a significant human and financial drain, so responsibilities should be drawn early. Assume for example that the firm is converting from Siemens to Avaya technology. Understanding what each user needs to have programmed into the station and how that relates to the new system (station programming) is a big job. The translation is not straightforward and requires considerable manual effort.
- Do the PBX and related systems support hoteling — a setup that allows employees infrequently in the office to direct the system (usually via a kiosk) to ring their assigned number at the telephone they are using that day?

Having discussed premise equipment or alternatives, we can turn to public network access and related cost reduction.

Local Access Strategies

Short of running a naked wire from one building to another, communications with the rest of the world require a local connection to one or more telco POPs (points of presence). The technologies, billing plans, and architectural design of these local connections will strongly affect the total costs of telecommunications. The following sections outline sources of savings and long-term strategies for linking the workplace to common carrier facilities. Some concerns, such as unused circuits and excess capacity, were addressed in Chapter 3.

Maximizing Traditional Access Circuits

The number of trunks (i.e., a phone line available for a single call) should equal some fraction of the total number of users. Many organizations use ten percent as a starting point, assuming that at any given time, approximately ten percent of the employees will be using the telephone. If trunk usage is plotted over the course of a business day, the resulting curve is usually a "butterfly," with the maximum number of calls occurring in late morning and early afternoon (see Exhibit 4).

Miscellaneous Technical Solutions

Reducing demand for communications necessarily reduces telecommunications costs. By properly managing trunking capacity and reviewing the

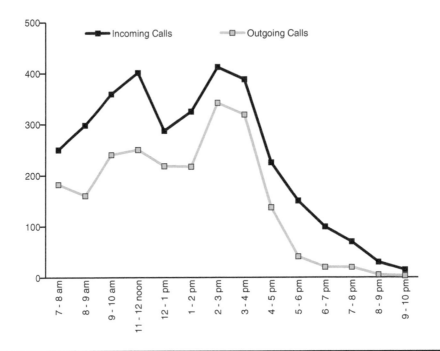

Exhibit 4.   Example of Trunking Usage by Hour

reports available, the telecom manager can avoid excess capacity. The industry standard is "P.01," which means one call blocked (due to lack of trunk availability) for every one hundred calls. For some critical groups/individuals within the organization, such as safety personnel or senior executives, trunks can be dedicated to lines, making them "nonblocking." It is not economical or necessary to do this for all employees at a location.

Another source of excess costs is inadequate/excessive trunking for long-distance calls. Typically, certain trunks are dedicated to local calls and others to long-distance (both inbound and outbound). If all the long-distance lines are busy, the PBX, fixated with getting calls through, will use every available means to do so, including using local trunks for long-distance calls. The local provider is all too happy to carry long-distance calls, charging "standard" rates that are likely far above rates negotiated with the long-distance carrier. A sudden rise in charges from the local telephone company may reflect long-distance calls that are redirected from busy IXC (interexchange carrier) trunks. Again, the telecom manager should be reviewing usage to ensure that the right balance is maintained.

Digital Subscriber Line (DSL)

DSL (digital subscriber line), cable modem, and other high-speed technologies should be considered in areas where they are available.

DSL services are growing rapidly as local access carriers continue to install DSL access modules ("DSLAMs"). Most service providers break down their DSL services into three main categories: residential, SOHO (small office/home office), and enterprise. ADSL (asymmetric DSL, where up- and downstream speeds differ) appears to be the offering of choice for residential customers, while SDSL (symmetric DSL) is usually marketed to businesses because it has T1 or more speeds both ways.

Depending on the geographic location, installation time for DSL circuits can range anywhere from one week to over ten weeks. Quality of service (QoS) is an issue. Most service providers, because of the multiple risk factors, do not guarantee QoS with DSL service. For example, DSL can potentially be unavailable for a few hours or longer. However, unlike cable, DSL provides consistent bandwidth to the user and does not depend on how many other (unrelated) customers are using the service at any one time.

DSL, more prevalent than cable modem, has a number of potential advantages:

- Internet access for smaller offices.
- Backup Internet connection. Many organizations implement DSL as a failover device because it is so inexpensive compared to a T1. Although costs vary by provider, a business typically pays at least twice as much and sometimes three or four times as much every month for a T1 line as it would for DSL. In other words, for the cost of a single T1 1.54 Mbps connection, three 1.1 Mbps DSL connections can be supported.
- Primary connection for the Internet and data services (note: many organizations do not consider DSL sufficiently robust for their primary link, but if cost is the primary consideration, DSL will provide the necessary functionality).
- Decreased installation charges. Typically, installation costs for a T1 line are three to four times as much as installation and setup of DSL services. DSL uses traditional telephone lines as opposed to T1, which requires installation of special (conditioned) lines.
- Multiple pricing categories. Depending on needs, more or less bandwidth (line speeds) can be purchased (if available in the area).
- Combined data and voice services over a single connection (for small offices). Although not yet widely deployed, Voice-over-DSL (VoDSL), shown in Exhibit 5, is starting to be implemented in selected locations in the United States, such as Santa Clara, California, and Boston, Massachusetts. By combining multiple voice channels and data on one copper wire, local telecom companies can offer a competitive package to small businesses that need only Internet access and a few voice lines. Jim Greenberg, chief architect at Rhythms NetConnections Inc., estimates small businesses could save about 30 to 40 percent on additional voice lines and get it all from one company.
- An "always-on" technology, unlike ISDN, which requires a sign-on.
- More secure than cable modem, because the bandwidth is not shared with other users (due to DSL's copper wire to the user legacy ar architecture — dedicated to a single user).

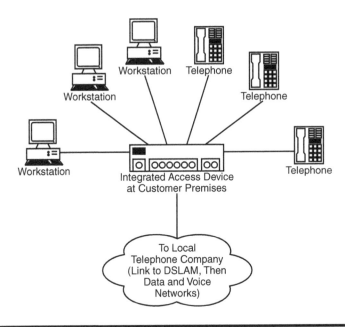

Exhibit 5. Voice-over-DSL (VoDSL)

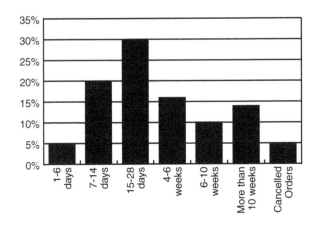

Exhibit 6. DSL Installation History: Percentage of Installs over Time of Order
(Courtesy of DSLreports.com, February, 2001.)

Disadvantages include:

■ DSL installation, while generally faster and cheaper than T1 or T3 installation, may experience technical problems. DSL runs over lines designed prior to World War I. It was originally intended to carry only miniscule traffic. For such a scrawny system to shoulder mountains of Internet data is akin to one writer's quip about a dog walking on its hind legs — "It is not done well, but you are surprised to find it done at all." Exhibit 6 shows an analysis of the percent of DSL installations completed versus time required for the install.

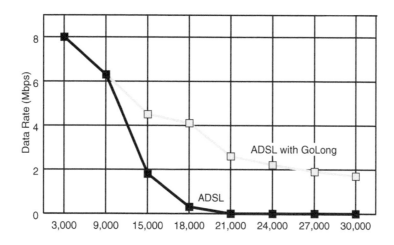

Exhibit 7. Distance versus Bandwidth for DSL (Courtesy of Symmetricom.)

- Multiple parties are involved. Typically, when a local telephone company, ISP, and possibly a DSL provisioning company are involved at some point in providing the service, the potential for billing errors and increased repair time is greater.
- DSL bandwidth varies considerably. The distance from the POP largely determines the bandwidth that is available to a DSL customer. The distance limitation is usually considered 18,000 feet but "loop extenders" from companies like Symmetricom can extend the distance considerably. Exhibit 7 shows the relationship between bandwidth and distance from the central office.

Although pricing (see Exhibit 8) will undoubtedly change by the time this book is in print, using data from selected providers shows the general trends in monthly pricing for DSL services:

Cable Modems

Primarily targeted at the SOHO and residential markets, cable modems offer the most competitive pricing for the *potential* bandwidth it can deliver. Under ideal conditions, cable modems can provide 10 to 30 Mbps downstream and 128 Kbps to 10 Mbps upstream.

Over-subscription, the *bete noire* of cable promoters, whacks this ideal down several notches. The total available bandwidth is shared among users in a neighborhood as if they were on a LAN. Hence, the available bandwidth for individual users will vary according to the number of users on the system at any given time, how much they are uploading/downloading, and the type of cable modem[5] being used.

Security on cable systems is notoriously weak. Although security standards exist, such as DOCSIS,[6] these deal primarily with provider, not user issues. Every user on a segment of the cable can see the others' traffic. While

Exhibit 8.   Pricing from Selected DSL Providers

Provider	Speed	Approximate Monthly Cost
Verizon	768K down — 128K up	$35
	384K down — 384K up	$52
	768K down — 768K up	$60
	1.5M down — 768K up	$155
Covad	ADSL (residential only)	$55
	SDSL[a] 144K both directions	$115
	SDSL 384K both directions	$155
	SDSL 768K both directions	$260
	SDSL 1M both directions	$400
MCI WorldCom	ADSL 608K down — 128K up	$40
	SDSL 128K both directions	$160
	SDSL 384K both directions	$200
	SDSL 768K both directions	$400
	SDSL 1M both directions	$500
	T1 1.54M both directions (for comparison purposes; not a DSL product)	$1,500

[a] Symmetrical digital subscriber line. Provides digital bandwidth up to 2.3 Mbps both ways (up and down). It can function over distances up to 10,000 feet.

applications (such as credit card transactions using SSL) can be secured, general traffic over cable is not secure.

Exhibit 9 shows the relative costs of cable modem, DSL, and T1 transmission methods. If cost is the only issue, cable is the clear winner.

Other Technology Cost Savings

There are a number of technology solutions that can reduce overall organizational costs, even if they add slightly to the telecom budget. Call centers provide one such solution.

Call Centers

Although call centers would be an obvious choice for heavy massing of technological firepower, many organizations still rely too heavily on human agents to do work that could be done by computers and telephone systems. Examples include:

■ *Predictive dialers.* Anathema to families that enjoy a quiet dinner together without telemarketer interruption, predictive dialers allow agents to call efficiently. Not only does the predictive dialer actually make the call, but,

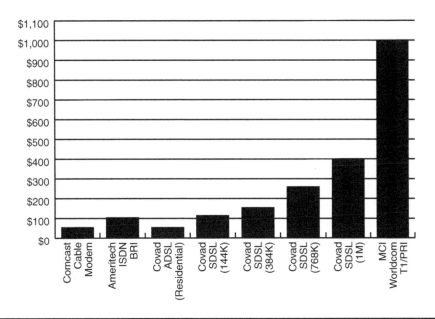

Exhibit 9.   Relative Costs of Broadband Transmission Methods

according to Richard Grigonis (*Computer Telephony Encyclopedia*), it "uses complex mathematical algorithms that consider, in real-time, the number of available phone lines, the number of available operators, the length of an average conversation and the average time operators need between calls, and constantly adjusts their dialing rates based on these factors." Also, the best predictive dialers screen out calls where there is no answer or those that are answered by an answering machine. Most of the time, the calling agent hears a quiet "zip" in the headphone and a live person is then on the line. While manual dialing may result in 15 to 20 minutes of productive calling time per hour, predictive dialers allow agents to productively talk 40 to 57 minutes per hour. Given that call center agents are paid between $12 and $20 per hour (as well as incentives), any device that makes them more efficient is likely worth the investment. It is interesting to note that, in the eternal war between "push" or outbound call centers and potential customers, technology solutions are found on both sides. Telemarketer "zappers" are now sold that intercept telemarketing calls. In Texas, some 77,000 households have signed up for a blocking service since the law went into effect on January 1, 2001.

■ *Call center workforce management software.* Although scheduling agents via software would seem to be a "nice to have," akin to a deluxe PDA, it strongly affects call center costs. Beyond a certain number of agents, it becomes difficult to mentally juggle schedules, demand, holidays, incentives, shifts, etc. One of the highest expense items is overtime; without an automated system for scheduling and reporting, absenteeism and overtime will climb to unacceptable levels (for mid- to large-sized call centers). Steven J. Cain, Gartner Group's Call Center Benchmarking Practice Research

Director, says that, "When you consider that, in some industries, contact center turnover reaches as high as 50 percent, there is significant opportunity to reduce turnover, building an experienced and tenured agent base to deliver the highest quality customer interactions while minimizing the expense of recruiting, training and productivity shortfalls while getting up to speed."

■ *Interactive voice response (VR) systems.* The familiar "press 1 for account balances, press 2 to transfer funds,..." is the public face of interactive voice response technology. Some call centers shun IVR systems because of the acknowledged public preference for human interaction. This philosophy should be reconsidered in some cases. For example, is it better to staff from 7 a.m. until 10 p.m. and then leave a message for the customer to "call back during business hours" or to have an IVR after-hours that provides the customer with some useful information. Second, as the public becomes more familiar with IVR, there are situations where non-human interaction is faster and preferred. For example, when people call about booking reservations for deluxe resorts, they want to talk to someone and ask multiple questions. However, if they must cancel those reservations, they merely want to cancel — why take the time to explain? In this case, the transaction can be handled without agent contact, saving money for the company and time for the customer.

Summary

How communications systems are constructed and linked to the public network strongly affects long-term costs. Choose the right equipment and the right access, and your organization can save thousands per month. A corollary to this common-sense rule is that there is no linear relationship between features and price. Like personal bills at home, each component must be individually examined and reviewed in light of market developments.

Notes

1. Sometimes called "next-generation" PBXs or "one-wire wonders," alluding to the elimination of duplicate wiring for data and voice.
2. Unified messaging applications feature voicemail, fax, e-mail, and even video clips in the same desktop inbox. They can be routed, saved, and translated from one medium to another (e.g., e-mail can be "read" by a text to voice translator).
3. In the same way that the indigenous peoples of the Arctic have multiple names for snow, because it looms so large in their lives, telephony people have many names for telephones: stations, handsets, voice terminals, extensions, among others.
4. Time division multiplexing. We have avoided a technical discussion about the origins of the TDM architecture here. From a cost perspective, the reader should associate "TDM" with traditional, usually proprietary, large PBX technology.
5. For bandwidth-intensive applications, the type of cable modem can have a major impact. Users will notice a visibly faster response when older cable modems are replaced.
6. Data over cable service interface specifications.

Chapter 6

Satellite Communications: Large Savings for the Right Applications

Satellites play important roles in finance, business, government, and international trade. For example, trillions of dollars are transferred over satellite links, airline reservations systems depend on them for a percentage of their business, and over 50 percent of all overseas communications is satellite based. Roughly 200 commercial satellites provide direct broadcast, GPS (global positioning systems), domestic/international linkages, and maritime and defense communications.

Much of satellite technology is mature. Some consider traditional satellite communications, such as VSAT (very small aperture terminal), to be obsolescent and moving to the end of its life cycle. The economic reality suggests a different view. Satellite communications costs for many applications are often one half to one third that of a comparable terrestrial network. Many of the installed satellite networks are relatively new and will be used as either primary or backup communications for the next five years or more. Finally, satellites can provide Internet services that are not economically viable using any other transmission medium.

A cursory analysis shows that satellite systems share a mixed risk profile. They are not usually subject to construction cable cuts or the vagaries of the public telephone network. On the other hand, the centralization of communications going through a single satellite exposes the entire network to downtime unless careful recovery planning has been implemented.

In order to put specific benefits and risks in context, the following section provides background information for the common satellite systems in use today.

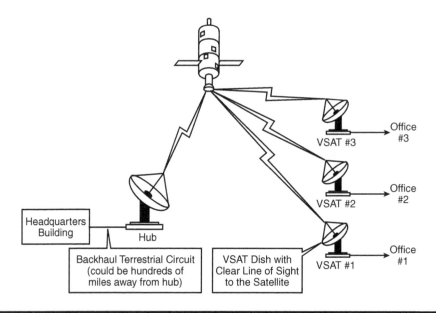

Exhibit 1. Geosynchronous Satellite System

Basic Satellite Technology

There are two primary satellite systems in use today: geostationary and low earth orbit (a third, middle earth orbit, is less common). Low earth orbit (LEO), due to its proximity to the Earth's surface, requires considerably less power for transmission, both to and from the satellite. Commercial enterprises, such as Teledesic, intend to use LEO systems to provide high-bandwidth Internet connectivity to areas of the world not readily serviced by terrestrial networks. LEOs can also be used for handheld mobile telephones in any location where there is a reasonably direct line-of-sight to the satellite.

This chapter focuses on geostationary satellite systems. Satellites in geostationary orbit appear to move only slightly in the sky, hence the term "geostationary." Flying high at 35,800 kilometers (22,000 miles), a geostationary satellite orbits the Earth in the same amount of time it takes the Earth to revolve once. From Earth, therefore, the satellite appears to be stationary, always above the same area of the Earth. The area to which it can transmit is called the satellite footprint.

Exhibit 1 is a simplified diagram of a typical geostationary satellite system using small receiving (end user) VSAT dishes.

Advantages of geosynchronous satellite systems include:

- *Quick to deploy at field locations.* Once agreements with the satellite and hub providers are in place, additional VSAT dishes can be installed within a few days. Terrestrial lines, on the other hand, can require months to install, depending on the carrier, location, and other logistical factors. Some organizations have mounted VSAT dishes on flatbed trucks so that they can be set up quickly; they can then be moved to another location.

Exhibit 2.   Sample Costs to Transfer *Internationally* 600 Megabytes to 10,000 Subscribers

Technology	Transmission Rate (Mbps)	Transfer Time	Transfer Cost (U.S. $)
Telephone line	0.04	2 days	1,500,000
ISDN	0.12	12 hours	100,000
ADSL	1.5	1 hour	20,000
Cable	4	20 minutes	1,200
Satellite	40	2 minutes	25

- *Only practical solution for very remote locations.* For example, in the natural gas industry, pipelines must be run across thousands of miles, some parts of which may be ten to hundreds of miles away from the nearest telecom point of presence (POP). In those cases, satellite transmission of technical data (e.g., pipeline pressure and flow) is the only reasonable way to monitor the pipeline. Similarly, some rural locations cannot get terrestrial services (other than ordinary telephone service) because their usage does not justify the high investment required by the local telephone company to run the conduit.
- *Low cost relative to terrestrial lines.* In many cases, satellite communications will cost one third to one half that of its terrestrial counterpart. Exhibit 2 gives example costs of shipping a large file internationally using various technologies. One-way video, in particular, is drastically less expensive than alternatives because of the satellite's ability to carry broadband transmissions. Sometimes, satellite communications are used solely as a backup in case terrestrial lines are down.
- *Works when parts of the public network are down.* As recent world events have reminded us, a temporary slowdown in the public switched telephone network is often a consequence of high-visibility disasters. Telephone lines and cellular phones (which ultimately use terrestrial lines) become swamped. In contrast, VSAT dishes need only electrical power and a clear line of sight with the appropriate satellite. They are unaffected by congestion on the public network or any circuit/equipment breakdown in the POP itself. The one exception to satellite independence from terrestrial carriers is the need to use conventional lines for the backhaul[1] circuit.
- *Capable of efficiently multicasting text, images, video, and audio over large geographic areas.* In contrast, multicasting via terrestrial lines often equates to repeated transmissions of the same information, wasting valuable bandwidth. Note that multicasting is a one-way, one-to-many medium.
- *Bypasses Internet congestion points* when used for Internet transmissions.
- *Only one or two providers needed for end-to-end international communications.* For terrestrial leased line management of international connections, each individual carrier usually manages its segment of the link. A satellite connection, on the other hand, may use only a single supplier who is responsible for the quality and management of the international backbone connection.

Disadvantages include:

- *Weather interference.* Bad weather (discussed in detail later) can disrupt satellite communications. Heavy rains, ice storms, and even the vernal/autumnal equinox can sometimes disrupt the signal.
- *Propagation delay.* Geosynchronous satellites must be positioned so high above the Earth that even traveling near the speed of light, signals do not instantly traverse from end to end. Voice carried over geosynchronous[2] satellite, while still used for mobile communications such as Inmarsat satellite phones, is considered inferior to landlines for routine business communications. The one-half to one-second delay is disconcerting for most individuals who are used to the full duplex mode of the traditional public network (i.e., both parties can talk at the same time, without delay). From a data communications perspective, this limitation makes both Internet access and highly interactive applications unsuitable for routine use over satellite links.
- *Limited two-way bandwidth.* While video "downlinks" from the satellite are carried over a large bandwidth, uplinks (from the VSAT dish to the satellite) are narrowband (typically 64 to 256 kbps).

Risk Areas for Satellite Communications

Experienced VSAT technicians are often loath to quote satellite reliability above 99 percent availability. Key risk factors for service interruption include the following:

- *Rain fade.* Normal or even reasonably heavy rain will not necessarily disrupt communications. However, a heavy downpour can weaken the signal so much that transmission stops. The hub operator can adjust power on the hub to a certain extent but at some level of rain, nothing more can be done. As expected, some areas of the United States are far more susceptible to rain fade than others (e.g., some parts of Florida).
- *Satellite malfunction — fuel shortage.* While any number of destructive elements, such as meteorites, can disable satellites, they are most commonly rendered useless because of fuel shortages. A geosynchronous satellite must necessarily stay within tight limits of position in the sky because all VSAT dishes must be fine-tuned for direction to ensure a strong signal. And because satellites naturally tend to wander in an elliptical path, they must be constantly homed to the correct position in space via small thrusters mounted at appropriate locations around the outside surface. These thrusters require fuel; when there is no more fuel, ground control is unable to keep the satellite on target and it drifts away, thus becoming useless for communications. There is a story about the early days of commercial satellites in which technicians new to satellite management used a mouse hooked to a control unit to position the "bird." Apparently, it was so much fun that they moved it around too much, depleting its fuel and rendering a multimillion-dollar satellite useless.

- *Satellite malfunction— transponders.* Transponders receive signals on the uplink, translate them to the downlink frequency, and amplify them for retransmission to Earth. Transponders can and do fail. Because there are multiple transponders in a satellite, the failure of a single transponder does not necessarily mean the end of the satellite's life. However, if an organization's communications are going through the failed transponder, the result is the same as if the satellite had been knocked out — that is, no service for that organization.

- *Ice and snow.* If VSAT dishes are not properly heated or enclosed in a radome (special purpose plastic cover), they may not receive and transmit a sufficiently strong signal to function.

- *Lightning and power surges.* Engineers working day-to-day on VSATs generally agree that the most frequent reason for breakdown of the dishes is electrical.

- *Relatively short mean time to failure.* VSAT equipment stays hot and wears out relatively quickly.

- *Frequency conflicts.* Although the FCC controls the frequencies used, occasionally a technician will set up a VSAT incorrectly, resulting in interference. The solution is for all parties to return to their assigned frequencies to avoid interference.

- *Sun transit errors.* When the main beam of an Earth station receiving antenna is in a straight line with the sun, significantly larger noise will occur, sometimes temporarily stopping communications. These errors are more likely to occur during the fall and spring equinox. Sun transit problems are far more likely to occur with the older, type I VSAT dishes. Type II and III dishes are smaller in diameter and less sensitive to concentration of noise.

- *Temporary "commandeering" of frequencies by government authorities.* For purposes of safety and security, government authorities or military personnel may temporarily take over certain frequencies in a geographical area. These are lawful actions but in some cases organizations using those frequencies are not timely notified and spend considerable time researching the cause of the downtime.

Controls to Improve Resiliency, Reliability and Security

Good practices can mitigate some of the risks associated with satellite communications. Firms that have all their offices or plants linked via satellite/ VSATs should carefully consider the exposures. For example, one paging company suffered customer ill-will and economic loss because the satellite it was using to relay pages stopped functioning. The following description of a satellite breakdown and its consequences is courtesy of *911 magazine* (August 1998, www.9–1–1magazine.com):

The Day the Pagers Went Silent

When the PanAmSat Galaxy IV Communications Satellite got knocked out of its orbit for a couple of days in May, it also knocked out the majority of pager communications in the United States.

Launched in 1993, the $250 million HS-601 spacecraft stopped relaying pager messages, television news feeds, and all sorts of broadcast data communications around 6PM PST on May 19th when the satellite's onboard control system as well as a backup switch failed and it rotated out of its proper position. PanAmSat, which owns the satellite, scrambled to establish communications with the Galaxy 4, finally re-establishing its position on the evening of the 20th. That affected thousands of emergency communications centers nationwide, which depend on pagers to notify responders and senior staff of emergencies...

"I would hope that in the future, this type of failure will be automatically corrected by electronic or computer means without having to manually redirect antennas or reprogram computers," said Miami's Charles Manetta. "This is how many telephone failures are corrected and are transparent to the end user. Time will tell."

The failure was not without irony. The *Phoenix Disaster Recovery Newsletter* reported:

For several hours after the spacecraft failure, the president of PanAmSat tried desperately to get in touch with Hughes' technical team in charge of engineering for Galaxy 4. After more than 3 hours, he finally contacted GM senior management (owner of Hughes, of course) by telephone. He said he'd been trying to contact Hughes' techno geeks for hours. "Why," he demanded, "didn't your people respond to my pages?"

Had a contingency plan been in place (including alternate satellite), resumption of service would have been quicker. Following are some of the most common control and security measures employed for VSAT satellite systems:

- *Change control.* Both the remote VSAT dishes and the central hub are attached to a myriad of software and hardware support systems. For example, central hub operators, including providers such as Hughes Global Services and Gilat Satellite Networks Ltd., must be extremely careful with the software that controls repositioning of the hub dish. Otherwise, an error could cause the signal to become so attenuated that communication would stop. Of course, the usual communications infrastructure, including hubs, routers, and network management software, should also be included in change control. Occasionally, perhaps once per year or every six months, hubs need to be brought down for maintenance (physical and software upgrades). This schedule should be published well in advance.
- *Equipment redundancy.* Spares for critical equipment such as the IP gateway (links the organization's LAN/WAN to the satellite system), specialized modems, encryption boxes, and other satellite-specific devices should be available and periodically tested.
- *Backhaul redundancy.* The backhaul circuit is usually a terrestrial communications link, such as a T1 or Frame Relay circuit, that connects one or more central locations to the satellite hub. If this link is cut by a backhoe

or loses function for some other reason, communication is lost. Hence, a duplicate circuit, perhaps from a different long-distance provider, but at least in a different conduit, is required.

- *Power.* For redundant equipment that is on hot standby, a separate power source provides protection from power supply failure. For example, a dual 250W hot-swap redundant power supply may be required for some devices.
- *Backup arrangements.* Satellites are expensive. The launch alone is typically $50 to $400 million, with costs further exacerbated by occasional launch failures. As a result, satellite transponder space is at a premium. Organizations relying on satellite communications for critical business functions cannot assume that they can "throw money" at the satellite vendors and get backup service quickly. Much of the capacity is booked months, even years in advance. Spare capacity should be obtained in advance of need.
- *Disaster recovery planning and testing.* In addition to negotiating with their satellite provider for backup capacity, organizations need to carefully design their response to a satellite failure. VSAT dishes will most likely need to be repositioned in every office or plant using the service. The whole reason for having a satellite remain geo-stationary is that the field dishes can be set and locked to look at a specific point in the sky (azimuth[3] and elevation specifications). Practically, it may take weeks for a large network of VSATs to get repositioned and correctly adjusted, because a trained technician must do the work.
- *Service level agreements (SLAs).* Service level agreements should be established for the hub operator, satellite service, and dish maintenance vendor. Frequently, the hub operator and dish maintenance vendor (for field locations) is the same provider. SLAs are particularly important for satellite failure because that is the most difficult step in recovery. If, for example, backup transponder space has been purchased on the same satellite, then the SLA should state how long it will take to transition operations. From the perspective of the field office or plant, what is the response time for dish or RF (radio frequency) equipment problems? Exhibit 3 summarizes key issues to be addressed in satellite service level agreements.
- *Capacity planning.* While the downlink bandwidth (satellite to VSAT dish) can be quite large, the uplink is often no more than could be expected from a terrestrial modem and sometimes less. As more VSATs are added, the uplink capacity of the system will degrade unless more "in-routes" or uplink bandwidth is added. If an organization has specific bandwidth needs that are highly likely to occur, it should consider purchasing extra transponder space so that there is no delay when the need arises.
- *Network Management System.* Components of the satellite communications system should be SNMP (Simple Network Management Protocol) addressable so they can be monitored along with the rest of the organization's communications infrastructure.
- *Physical/electrical protection.* For VSAT dishes, a lightning arrestor and surge arrestor are *de rigueur*. Trees, bushes, and other obstructions can interfere with the line-of-sight. Often when the dish is installed, adjacent trees are small but with growth they steadily decrease the signal strength. Access to the facility should be restricted as well.

**Exhibit 3.   Example Service Level Agreement Categories
for Satellite Operations**

Component	Example/Comments
Overall communications uptime, end-to-end	Average uptime of 99.78 percent across the whole end-to-end link from the New York Internet node all the way up to the customer's local premises.
Time to restore services:	
Terrestrial equipment failure	Standard time to repair VSAT dish is 8 h; emergency repair is 4 h from time of incident notification.
Backhaul interruptions	Carrier will restore services within 4 h of notification by a traffic reroute or other means.
Hub failure	Provider will restore services with 4 h of notification. After 8 h downtime, provider will initiate disaster recovery plan and restore service within 48 h.
Notification of downtime	Provider will notify customer no less than 30 days prior to any scheduled hub downtime. Customer will be notified immediately if unscheduled maintenance is required.
Spare parts inventory	Provider will maintain spare parts for VSATs within a reasonable distance of customer operations (no more than 50 miles from premises).

- *Expertise.* For those firms with enough VSATs to justify owning their own hub — an investment in excess of $1 million — highly skilled technicians are required. Backup personnel (perhaps including contractors) should be available.
- *Spare parts.* Particularly for hub operators, spare parts will prevent delays in operations.
- *Documentation.* As in other complex systems, documentation of frequencies, sites, IDs, network schematics, etc. is important. Firms operating their own hubs need to pay particular attention to documentation because of the inevitable drift toward technical uniqueness.

Other Concerns

Propagation delay is one of the primary drawbacks to geosynchronous satellite usage. For organizations that do batch transmissions (mostly one way at a time), delay is acceptable and largely irrelevant. However, for interactive work, such as entering accounts payable information on a client/server architecture, the one- to two-second delay is annoying.

With effort, interactive applications can be designed to send only essential data over the satellite link. The concern is whether expectations are managed for any new applications. Most programmers and systems designers are accustomed to large bandwidth. Hence, most off-the-shelf applications, not specifically designed for satellite links, will be sluggish or even unacceptably slow.

When reviewing the economics of satellite communications, contractual agreements should be reviewed carefully. For example, if the organization views satellite as an interim solution, are "ramp-down" provisions included in the contract so that costs are not excessive as the shutdown date nears? On the other hand, if the organization starts out with only a few VSAT locations but intends to rapidly deploy the rest of the network, a ramp-up clause should be included so that unit charges during the early phase of implementation are not excessive.

The Future for Satellite Technology

Although traditional VSAT technology, with its minimal uplink bandwidth, is not appropriate for some organizations, the newer systems in development should be reviewed by network architects. For example, Hughes' new system under development, the Spaceway system, is expected to provide a variety of low-cost broadband services with small satellite dishes, with data rates ranging from 512 kbps upstream and up to 30 Mbps downstream. Applications will include Internet access (with a strong multimedia component) to LAN/WAN solutions for work-at-home employees, SOHOs,[4] and large organizations.

Hughes' system includes full mesh point-to-point and multicast communications architecture. This allows the development of high bandwidth peer-to-peer applications, such as file sharing, distributed databases, and decentralized content distribution.

The availability of reasonably fast Internet links in rural areas around the world could significantly change the business dynamic of many firms. While the media continually laments the lack of bandwidth, the most serious deficiency of the Internet is actually the lack of geographic coverage.

Another alternative architecture is a hybrid system that uses satellite transmissions for downlink and terrestrial for uplink (currently used to provide Internet access to areas with no other broadband availability). Because satellites are large (many tons), they have power plants that allow megabit-per-second downloads of video, software upgrades, and other information. The terrestrial link in this asymmetric data access scheme provides for less latency (delay) for the user response. Most applications, as is the case with home Internet users, consume far more download bandwidth than upload bandwidth.

The technology of caching will be increasingly used for Internet services. Caching takes recently retrieved information, copies it, and places it on a server close to the consumer. This process allows users to access popular Internet data quickly because it is physically located much closer to the user. The more users are associated with a cache, the more the benefit because there will be a higher likelihood that a requested file will be in the cache. This could potentially speed the deployment of international intranets for global organizations. Caching is relevant to satellite transmissions because it reduces demand for repetitive uplinks from the hub for frequently used pages.

Summary

Any organization that uses satellite technology faces the risk of temporarily losing an entire communication network. On the other hand, satellite technology offers a plethora of economic benefits if the fit is proper, based on the organization's business needs. Satellite links are usually less expensive than Frame Relay and other leased landlines. In some cases, they are the only alternative.

Notes

1. Backhaul has come to designate terrestrial communications, such as a T1 or Frame Relay, that link a hub or distribution center with the corporate office(s). For example, if the hub where the satellite signal is received resides in Detroit, Michigan, and the corporate headquarters are in Houston, Texas, then a line is necessary to connect the two locations.
2. The distinction between geosynchronous and LEO (low earth orbit) is stressed here due to the significant differences in capabilities.
3. Azimuth relates to the left-to-right positioning of the VSAT dish. Elevation is the angular distance of the satellite above the horizon in relation to the dish. Thus, elevation relates to the up-and-down positioning of the receiving dish.
4. Small office/home office.

Chapter 7

Telecommunications Security

Trust not him with your secrets, who, when left alone in the room, turns over your papers.

— Johann Kaspar Lavater, 1788

Security breakdowns in either voice or data communications nearly always result in financial losses and many times public embarrassment. We include this chapter on security because of the near certainty that loose controls will result in at least moderate losses over time.

Losses from intentional security breaches fall into the following categories:

- Theft of long-distance services
- Business losses (actual or opportunity costs) due to disclosure of confidential or proprietary information or service disruptions
- Destructive acts (such as deletion of critical customer information)

Security is now much in the news and the discussion in this chapter merely posits the broad outlines of effective defense. However, as Woody Allen quipped, "80 percent of success is showing up." A lot can be gained by implementing basic controls.

Voice and Telephony Security

A company's vulnerability to threats varies by its size and business type. For example, businesses that frequently engage in intense international bidding may find themselves in competition with a government-owned organization. Because the government often owns the telephone company as well (PTT[1]), there is a temptation to "share" information by tapping the lines (all it takes is a butt set and knowing which trunks to tap into). While such occurrences are undoubtedly infrequent, they are a threat.

Toll fraud, on the other hand, is ubiquitous. Hackers use stolen calling cards to find a vulnerable PBX anywhere in the world and sell the number on the street (mostly for international calls). Poorly controlled voicemail options and DISA (direct inward system access) are excellent "hacker attractor" features. Medium-sized installations are preferred because they offer enough complexity and trunking to allow hackers to get into the system and run up the minutes before detection. Smaller key system sites do not have the capacity, and larger sites often (but not always!) have toll fraud detection systems (such as Telco Research or ISI Infortext's TSB TrunkWatch Service).

Two characteristics of the telephone system enhance the hacker's world of opportunity: (1) it is difficult to trace calls because they can be routed across many points in the system; and (2) hacking equipment is relatively cheap, consisting of a PC or even a dumb terminal hooked to a modem. Hackers (a.k.a. "phone phreaks") sometimes have specific PBX training. It could be a disgruntled PBX technician (working for an end-user organization or the vendor). In addition to their technical background, hackers share explicit information over the Internet (see www.phonelosers.org). These individuals have a large universe of opportunity; they hack for awhile on a voice system, find its vulnerabilities, and then wait for a major holiday and go in for the kill. Losses of $100,000 over four days are common. If holes in one PBX have been plugged, they go on to another. In some cases, they use a breach in one PBX to transfer to another, even less secure PBX.

The final category of security break, malicious pranks, gets inordinate attention from senior management — far beyond the economic damage usually incurred. For example, a voicemail greeting could be reprogrammed (just by guessing the password) to say, "Hello, this is Mr. John Doe, CEO of XYZ Company. I just want you to know that I would never personally use any of XYZ's products." Of course, not all changes are minor. A clever hacker who obtains control of the maintenance port can shut down all outgoing calls or change a routing table — there is no end to the damage if the maintenance port is compromised.

The following sections list practical steps that every organization should take to reduce the impact of security breaches.

Toll Fraud

Prevention of toll fraud requires unceasing vigilance. Hacking is frequent and can result in large losses. For example, NASA and the Drug Enforcement Agency have both been hacked for millions of dollars.[2] The basic steps for toll fraud prevention include the following:

Exhibit 1. Example of PBX Maintenance Port Protection Device (Uses Two-Factor Authentication) (Courtesy of CDI, www.commdevices.com.)

- *Protect the PBX maintenance port.* Use passwords of at least ten characters and change them monthly. This is the absolute minimum protection. Far better is to use a two-factor authentication system,[3] such as verification systems from Axent, CDI's Uniguard, or Avaya's ASG security gateway. Exhibit 1 illustrates a device used to control access to multiple ports, including the PBX. Such a device can be used to manage security for many devices.
- *Use common sense calling restrictions.* If your organization never makes calls to South America, restrict the calling patterns to eliminate that possibility. The telephone operators can be given a class of service that overrides that restriction on the chance that a legitimate call needs to be made to a restricted location. Calls can be restricted by time of day, day of the week, and location. For example, lobby area telephones should not generally have the ability to make long-distance telephone calls (or at least not international calls). If the organization does not do business on Sunday, restrict outgoing calls on that day. All "common area" telephones, such as those in lobbies, break areas, and conference rooms, may need to have after-hours restrictions. The mechanism for restricting functions on the PBX is the class of service. Exhibit 2 lists a typical scheme for class of service in a large office. Many organizations, much to their later regret, have allowed the technical staff to set class of service policy. Because the technical staff is oriented toward pleasing the user, there is often escalation over time in the number of users who have the most powerful class of service. In the absence of policy, if a vice president asks a switch technician to enable dial-tone capabilities from an international location, the switch technician will most likely comply with the request.
- *Use toll fraud insurance.* Some PBX vendors and most common carriers will provide toll fraud insurance, as long as basic control mechanisms (that they specify) are in place. Typically there will be a deductible ($5000 to $20,000) per loss, but at least coverage for large losses is available. The carriers have sophisticated monitoring programs that identify an organization's typical usage patterns and flag unexplained and rapid increases in volumes to particular destinations. Also, some international locations are far more likely to be called by hackers than others (actually, hackers typically sell the "service" to individuals on the street, who then tend to call certain locations more than others).

Exhibit 2.   Illustrative Class of Service Parameters

Feature	Risk Factor
Profile for an executive extension	
Internal calls	None
Local calls	None
Domestic long distance	High
International long distance	High
Automatic camp on busy	None
Always in privacy	None
Call forward external	High
Call forward internal	None
Camp on busy	None
Conference call	Medium
Control of station feature	High
Direct call pick up ("pick")	None
Direct trunk select	Medium
Executive override	Medium
No howler off-hook	None
Private call	None
Save and repeat	None
Station speed	None
System speed call	None
Trunk-to-trunk	High
Profile for a lobby extension	
Internal calls	None
Local calls	None
Call forward external	None
No howler off-hook	None
Station speed	None
System speed override	Low
System speed call	None

■ *It is prudent to keep an up-to-date contact list of those management personnel authorized to make decisions regarding long-distance services.* This list should be periodically sent to the vendor (carrier or PBX manufacturer) that is monitoring your traffic. For example, assume that your organization is attacked on a Saturday night. The monitoring service identifies hundreds of calls going to Bolivia and Columbia (countries with which you normally do not do business) and attempts to call a responsible party on your contact list. If they cannot reach someone in authority, they are hesitant to shut down all outgoing international business because you may have essential functions that require outgoing international calls.

■ *Put tight controls over tandem trunk calling (going into the PBX, then going to an outside line).* DISA — allowing someone to call in, get dial tone, then call out — should be prohibited unless there is some security system in place to control it (e.g., voice verification). Some organizations will allow calls into voicemail, and then a transfer to dial tone (using a password). Given the ease of password cracking techniques now available, this service to employees can be expensive indeed. Better to provide them with calling cards for business-related calls outside the office (or an 800 number to dial into the office). Sometimes, vendors set up a new PBX and voicemail system and leave backdoor passwords as well as voicemail-to-dial tone capabilities (with only a two-digit password). In smaller locations, the organization will be completely dependent on vendor expertise. When a hacking incident occurs, the maintenance vendor may accept the responsibility or may say that the customer never instructed them to eliminate DISA, etc. *Caveat emptor!*

■ *Periodically review forwarding of extensions to dial tone.* Any station forwarded to dial tone is "hacker bait."

■ *Educate your operators and employees to social engineering techniques.* One technique widely practiced is for a hacker to call someone and say, for example, "I'm from PAC Bell and we are testing your system for some reported problems. Would you please forward me to 9011 so we can complete our trace of the system?" Of course, this transfer gives them dial tone. Another scam is for someone dressed in a delivery company uniform to arrive at the receiving desk to deliver a package for "Mr. X." Mr. X is not there and the hacker asks to use the telephone to call his boss. Apparently, he is put on hold and then gets in an involved conversation with his boss about wrong directions, etc. What he is actually doing is dialing a local number that charges a high per-minute charge for services (e.g., $15 per minute); he then gets a kickback from the service provider.

■ *Immediately request your local exchange carrier to disallow any third-party charges to the main number.* Some prisoners, for example, will make long-distance calls and charge to any organization that allows third-party charges.

■ *Do not forget to periodically review your call accounting reports.* Are there calls to a location that your organization has no business reason to call? Some hackers will keep the volume of calls sufficiently small to stay below the radar screen of the long-distance carrier's monitoring algorithms. Sort down minutes called by location and also list single calls in descending order of cost. A quick review can spot problem areas — including some that are unrelated to toll fraud (e.g., "stuck" modems).

■ *Educate users on the vulnerability of calling card theft.* In some airports, "shoulder surfers" observe calling card numbers being keyed in and sell the numbers on the street as fast as possible. Using an 800 number to call back to the office reduces the frequency of calling card calls (as well as reducing the cost). Using a voice verification system to allow secure DISA (see discussion below) also decreases the need for card use. A user, in the interest of expediency, may occasionally give her card number to co-workers. Most carriers, when they detect multiple usage of the same calling card in widely separate geographic areas (e.g., Japan and the United States) within a short period of time, assume fraud. Ensure that all employees who need a card have one.

■ *Some organizations, concerned about potential misuse by their own employees, contractors, or temporary workers, use prepaid calling cards.* The advantage of this technique is that a stolen card number would be used to its limit and then no further charges will accrue. The disadvantages are that it allows for no internal accounting of what the card was used for and that sometimes the card is not fully used.

■ *Monitor your organization's fax-on-demand server.* To efficiently serve their customers, many firms will set up a fax-on-demand server that accepts a call from the public network and faxes requested information back to the caller. Hackers have recently begun to exploit this service in the following ways:
  – Repeatedly calling the fax-on-demand service, asking for faxes to be sent to a 900 or 976 number owned by the hacker (these area codes have a special surcharge associated with them). Of course, the information on the fax is not used, but the minutes accumulate and the calling party (i.e., the hacked party) is responsible for paying the toll.
  – Repeatedly calling a fax-on-demand service, merely to harass the organization by running up its long-distance bill.
  – Harassing individuals by sending the fax to a business or residence that did not request it (waking up people in the middle of the night, etc.).
  – One company was hit with over 2000 requests to send a long document to Israel, resulting in a $60,000 telephone bill.[4]
  – Techniques to detect and defend against fax-on-demand abuse include:
    ■ Check the fax system log (or call detail) for repetitive faxes to the same number.
    ■ Exclude all area codes where there is no reasonable expectation that the organization would do business.
    ■ Exclude area codes associated with high fraud incidence (e.g., 767 — Trinidad and Tobago; 868 — Dominican Republic).[5]
    ■ Monitor overall volume of faxes sent out.
    ■ Power off and on to clear the queue if it is obvious that the server has or is being attacked.
    ■ Monitor the fax server over the weekend (particularly long holiday weekends) because that is the favorite time for hackers to start their penetration.

■ *Make use of your organization's internal billing system.* It is easier to spot unusual activity if long-distance bills are broken down by department. Make the internal reports easy to read, with appropriate summary information (e.g., by international location called), to provide the organization with more eyes to watch for unusual activity.

■ *Use appropriate hardware/software monitoring and toll restricting tools.* Some features of these tools include:
  – Selectively allow or restrict specific telephone numbers and/or area codes.
  – Allow 0+ credit card access but restrict 0+ operator access.
  – Limit the duration of telephone calls in certain areas.
  – Restrict international toll access.

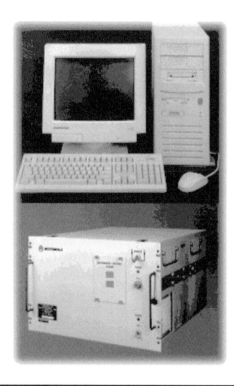

Exhibit 3. **Motorola End Crypto Unit, CI-13** (Courtesy of Motorola, Inc.)

- Provide for bypass codes.
- Report on a daily basis (sent via e-mail) any suspicious activity, based on predefined exception conditions.

Business Loss Due to Disclosure of Confidential Information

Some organizations have found their bids for projects coming in at just above the competition on a consistent basis. This could be due to coincidence or to unauthorized disclosure. It is always a concern when sensitive information is passed over wires or air space.

Following are some techniques for securing confidential voice transmissions:

■ Use a scrambling device such as SecureLogix Telewall, which has built-in encryption capability (the same device is required on both ends). The advantage of a trunk rather than handset-based approach is that the entire office or plant can be set up for encrypted conversations, assuming the other end (e.g., headquarters or a sister location) has a Telewall as well. The Motorola KG-95 also encrypts at the trunk level, unlike the older AT&T Surity 3600, which encrypts only from one handset to another. The Motorola product is shown in Exhibit 3. These devices, which enable point-to-point and multi-party encryption, protect the conversation from origin to destination (i.e., no intermediate points of clear conversation). Faxes can be protected as well. They typically have a secure/non-secure button that allows the telephone to be used in either mode, as required.

■ Use IP encryption if the voice conversation is converted to IP traffic before transmission beyond the premises. The Borderguard NetSentry devices, for example, use DES (Data Encryption Standard), 3DES (triple DES), and IDEA (International Data Encryption Algorithm) to scramble any data going across the wire. Note that with the increasing power of microchips, it is much easier for determined hackers (or governments) to break codes. The following quote, found on an Internet security page (http://www.jumbo. com/pages/utilities/dos/crypt/sfs110.zip.docs.htp), illustrates how quickly algorithms once thought secure have become as antiquated as iron safes:

> *Use of insecure algorithms designed by amateurs. This covers the algorithms used in the majority of commercial database, spreadsheet, and word processing programs such as Lotus 123, Lotus Symphony, Microsoft Excel, Microsoft Word, Paradox, Quattro Pro, WordPerfect, and many others. These systems are so simple to break that the author of at least one package which does so added several delay loops to his code simply to make it look as if there was actually some work involved.*

■ Use an enterprisewide dialing plan to ensure that all calls go through the least cost and least public route. Calls that go over leased lines (tie lines) are easier to secure than calls going over the public switched telephone network (PSTN). Encryption equipment can be placed at both ends and the voice traffic can be converted to IP. Typically, dialing plans are implemented to facilitate ease of use for employees as well as least-cost routing. However, they also increase (at least to some extent) security. A dialing plan is implemented by making changes to every PBX in the organization's network so the user dials the same number to reach an individual regardless of what location the call is made from. For example, if Mary Doe's number is 789-1234 and she is located in a Memphis, Tennessee office, then she can be reached from London or Sydney by dialing 789–1234 (with no preceding country codes, etc.). The PBX has all the logic built in to convert the numbers to the appropriate route. A dialing plan also has the side benefit of increasing contact between the telecom staffs of various locations, resulting in an exchange of security information.

Keep in mind that the U.S. Commerce Department as well as most international governments have significant regulations on the level of encryption used. The French government, in particular, has stringent laws against encrypting without permission.

Malicious Pranks

Many of the same controls listed for toll fraud will help reduce the exposure to destructive changes by hackers. Some basic prevention steps include:

- Force changes of voicemail passwords. Most current voicemail manufacturers maintain a history of changes so that a user cannot change his password to one number and then quickly change it back to the same number he has used for the past ten years.
- Force passwords to be at least eight digits.
- Identify unused mailboxes (sometimes used by drug dealers as an untraceable mailbox for transactions).
- Never allow dial tone to be accessible from voicemail.
- Implement a class of service program that allows employees or on-premise contractors to have only the features they need. For example, the ability to modify someone else's telephone features is obviously powerful and dangerous if misused — a hacker who gains access to a phone with that level class of service could significantly disrupt operations. Review class of service annually.

Using Security Tools to Offer More Services

Although our discussion of security to this point has been from a defensive perspective, there are a few operational enhancements that come out of a good security system. Some of these include:

- *Use of voice verification to allow DISA.* By enrolling employees who normally use calling cards for business (salespeople, traveling professionals, etc.) in a voice print authorization system, calling card costs can be significantly reduced. By use of an 800 number to call in to the PBX and allow DISA for an outgoing call (after verification), a traveler can obtain the same services at a cheaper rate. Although she would pay for the call two ways (into the PBX and out to another location), the cost of calling card calls is usually so high that the organization still reduces costs. In particular, the cost of calling card international calls and intraLATA calls are often well above 800 number rates. Exhibit 4 shows a payback analysis using fictitious but typical calling card and 800 number rates. Savings in calling cards alone can pay for the security device, since the payback shown in less than one year. Of course, the payback calculation shown in Exhibit 4 will vary considerably, depending on the number of calls via calling cards, the percentage of users who would be willing to go through the voice registration process, per-minute costs of long-distance and calling card usage, and cost of the verification equipment itself (e.g., Veritel's Voicecheck technology).
- *Access voicemail in areas of the world without touch-tone telephones.* Using voice-activated-only voicemail (with appropriate speaker voice recognition) allows rotary users to go through menus within voicemail.
- *Access special/confidential services.* For example, Parlance Corporation has a service called Employee Connector that allows an individual to list multiple phone, pager, cellular, etc. numbers. These numbers can be dialed by saying, for example, "Ms. Doe's vacation home" or "Mr. Smith's New York office." Having this information would be useful for executives and their administrative assistants but might be too sensitive for the general

Exhibit 4. Analysis of Potential Savings Using Voice Verification in Place of Calling Cards

Step #1: Calculate total costs of calling card usage

By number of calling card calls per month	$30,000
By minutes of use per month	$150,000
Blended cost per minute (average of interstate, intrastate, intraLATA)	$0.21
Setup cost per calling card call	$0.25
Cost per minute of calls to and from business locations ("mixed rate" — dedicated to switched location)	$0.09
Cost for minutes	$31,500
Cost for setup	$7500
Total calling card costs for the organization	$39,000

Step #2: Calculate voice verification costs using same assumptions

Cost of minutes — calling in via 800 number	$13,500
Add same number of minutes calling out	$13,500
Total voice verification costs using secure DISA (in and out)	$27,000
Monthly savings assuming 100 percent usage of voice verification	$12,000
Monthly savings if only 25 percent of calls are via voice verification	$3000

Step #3: Calculate payback

Representative cost of voice verification equipment	$20,000
Monthly savings (25 percent of calls use voice verification)	$3000
Payback period in months	6.7

Note: All costs above are for illustrative purposes only and do not represent actual prices from any carrier or vendor.

employee population. By front ending this service with a security device, it would be practical to use it. Executives would feel confident that only those with a need to know would have access.

PBX Firewall

In recent years, PBX firewall technology has been increasingly used by military installations, hospitals, energy firms, banks, and others to dramatically increase their control of telecommunications traffic. In some cases, the increased reporting available from the firewall has enabled them to reduce costs as well.

How does it work? A series of very fast, special-purpose computers sit between the "demarc" wiring from the telephone company and the PBX. These pizza box size computers have the unique ability to look inside the traffic on the telecommunications lines and apply predefined, logical rules. For example, the PBX firewall can:

■ Stop or log unauthorized modem traffic (e.g., individuals may set up PCAnywhere on their workstations as a convenience, unintentionally leaving a backdoor for hackers).

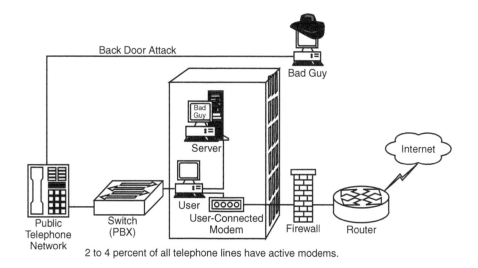

2 to 4 percent of all telephone lines have active modems.

Exhibit 5.   Standard Hacker Technique to Bypass IT Firewall

- Stop modem traffic detected on lines that are supposed to be for fax-only. The same event could also trigger logging-only or a page to security personnel.
- Stop voice traffic on fax lines after a certain time at night. Or limit calls to no more than three minutes (time to communicate with a distant party regarding fax status).
- Report on any lines not used in the past six months.
- Page or show exception reports when anyone in the organization calls a direct competitor.
- Disable any calls to or from ISPs not relevant to the organization's business.
- Provide special-purpose reporting on individual lines, odd usage, traffic between company locations, etc.

Exhibit 5 illustrates how hackers often thwart a strong IT firewall. Hackers, like most others, first look for the easy way in.

The PBX firewall, shown in Exhibit 6, sits between the demarc and the PBX, significantly lessening the likelihood of unauthorized intrusion (assuming the appropriate logic rules have been programmed).

Savings Potential Using PBX Firewall

The detailed information gathering and reporting available from the PBX firewall can potentially result in cost reduction. In a case study reported by Memorial Hermann Hospital in Houston, Texas, significant savings were obtained from:

- Elimination of unauthorized modem calls to ISPs, freeing up trunk lines for voice communications so that installation of new T1s could be eliminated or at least delayed.

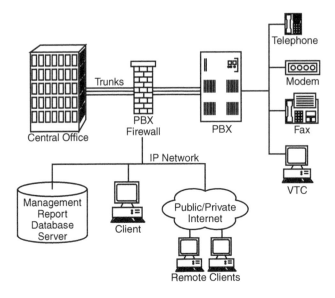

Central Office — Trunks — PBX Firewall — PBX — Telephone, Modem, Fax, VTC

IP Network — Management Report Database Server — Client — Public/Private Internet — Remote Clients

Exhibit 6.   Placement of PBX Firewall to Control Traffic

■ Replacement of higher-cost local access trunks with cost-effective tie trunks. By identifying that much of the PSTN traffic was between Memorial Hermann locations, it was easy to justify fixed cost tie lines that proved to be less-expensive, even in the short run. Three local access trunks costing a total of $4500 per month were replaced with three tie lines at $1050 per month, resulting in a savings of $3450 per month.

■ Reduction of full-time equivalent employee costs. The higher visibility of telecom information plus the ability to centrally monitor the entire Memorial Hermann enterprise resulted in decreased telecom FTEs per end user.

Another cost savings was reported by an East Coast banking enterprise. The telecom organization within the bank installed a PBX firewall for a few weeks at each of several locations. When the traffic patterns were recorded at one location, the firewall was moved to another site. After the round-robin process was complete, the bank had identified enough unused capacity to justify the firewall purchase several times over.

VoIP (Voice-over-IP) Security

With the proliferation of Voice-over-IP (voice–data convergence), new defenses are required. Because VoIP is a packet-based technology (i.e., in the data world), it must typically go through a firewall or outside the firewall. Either solution is less than desirable from a security perspective because it opens up the network to hacker attack on the VoIP gateway. One company, Quintum Technologies (www.quintum.com), has developed a solution (NATAccess) that gets around the problem, allowing only authorized traffic to pass through the firewall. According to Quintum Technologies, "It is now possible for systems administrators to deploy VoIP quickly, easily, and securely, without making major changes to their existing network infrastructure, or compromising their network integrity." Others will undoubtedly develop similar capabilities.

IP-based videoconferencing can have similar security concerns. In the January 2002 issue of *Internet Telephony,* Robert Vahid Hashermian notes that Microsoft's NetMeeting product has the following (rather technical) requirements, as noted in the Microsoft NetMeeting site:

> To establish outbound NetMeeting connections through a firewall, the firewall must be configured to do the following:
> - Pass through primary TCP connections on ports 389, 522, 1503, 1720, and 1731
> - Pass through secondary TCP and UDP connections on dynamically assigned ports (1024–65535)

The net effect of the above is to bypass the firewall and expose one's workstation to the world. This is an example of a generic risk that needs the attention of anyone planning widespread implementation of videoconferencing. The old circuit-switched (nailed up circuit) videoconferencing did not have these exposures.

Summary

As long as the communication system is connected to the public switched telephone network, it is vulnerable to attack. Reasonable security steps, including class of service policies, routine monitoring of traffic patterns by the organization and common carriers, installation of fraud detection equipment, employee education, periodic reviews by an outside party, and close communication with equipment vendors on security issues, can greatly reduce the frequency and severity of security breaks. Although not always easily quantified, good security practices and tools pay for themselves.

Notes

1. Post Telephone & Telegraph (telephone company usually owned by a country's government).
2. *Oregon Certified Public Accountant,* October 1994.
3. Two-factor authentication uses something you know (a password, user ID or PIN) and something you have (a token). The token may be software stored on the user's PC or in a separate handheld processor (a small device like a pocket calculator or pager). The token has within it the user's unique and secret key.
4. Web page from Epigraphx LLC, 965 Terminal Way, San Carlos, California 94070 (http://www.epigraphx.com/faxhacking.htm).
5. Web page from Epigraphx LLC, 965 Terminal Way, San Carlos, California 94070 (http://www.epigraphx.com/faxhacking.htm).

Chapter 8

Telecommunications Tax Minimization Strategies

The essential causes of Rome's decline lay in her people, her morals, her class struggle, her failing trade, her bureaucratic despotism, her stifling taxes....

— Will Durant, *The Story of Civilization*

To many, telecommunications services are abstract and complex. To others, the intricacies of federal, state, and local taxation belong to the "tax cognoscenti" alone, never to be understood by non-specialists. Certainly the intersection of these two — telecommunications and taxation — is a specialty within a specialty that is rarely reviewed. For some organizations, particularly larger ones, telecommunications tax strategy can save thousands, even millions of dollars. For example, one large computer manufacturer received a telecommunications Federal Excise Tax refund in excess of a million dollars.

Telecom tax savings strategies fall into the following broad categories:

- Partial refund of the Federal Excise Tax (FET)
- Reduction of sales and use taxes (state and local)
- Special tax considerations for industries and groups, such as:
  - Nonprofits
  - Common carriers
  - Telecommunications providers (e.g., special depreciation rules)

The following sections outline potential strategies in each category. Of course, readers looking to achieve tax reductions in these areas should consult tax specialists, where appropriate. For example, actually filing for the Federal Excise Tax refund is a somewhat convoluted process and probably not suitable for the non-specialist.[1]

The Telecom Federal Excise Tax Refund Opportunity

Ancient History

Section 4251 of the Internal Revenue Code (IRC) imposes a three percent excise tax on amounts paid for communications services, which include local telephone service and toll (long-distance) telephone service. Readers with a bent toward grammatical precision will be interested in the consequences of a single word, "and," that has resulted in hundreds of thousands of dollars of tax refunds for some large organizations. For much of the one-hundred-year history of the Bell Telephone Systems and its successors, long-distance calls were based on *both* the duration of the call (minutes) *and* the distance of the call (miles). Starting in the late 1960s, however, most contracts between carriers and their customers set pricing based only on duration, ignoring distance (at least for interstate calls). Because Section 4251 uses the critical word "and" in its description of the excise tax (i.e., communications services are taxed that are time and distance based), there is a reasonable argument that it does not apply (see Exhibit 1). The IRS, as one would expect, does not entirely agree. Without going into the minutiae of the legal and tax arguments, we can summarize the net result: (1) The IRS reviews the filing; (2) the client, assuming other appropriate conditions are met, receives a refund for a portion of the excise taxes paid over the past 36 months.

Details of the Opportunity

The IRS statutory language that governs the excise tax does not effectively address the following telecommunications services:

- Virtual private networks (VPNs)
- "Postalized" services
- Certain 800 (toll free) services

All the above fall within the purview of the potential refund because they are typically taxed at the standard three-percent rate (via carrier billing) but are only time, not distance sensitive. Because the statute of limitations is 36 months for filing excise tax refund claims, any organization looking to exercise this strategy should file as quickly as possible. Any potential refunds stop at the 37th month.

Firms can assess the opportunities for refunds by answering the following preliminary questions:

- Does the organization have any or all of the following contractual arrangements with telecommunications providers?
  - AT&T Contract Tariff or Tariff 12 or 15 w/Option
  - MCI Special Customer Arrangement
  - Sprint Custom Network Service Arrangement
  - Other carrier agreements with long-distance charges based on time only

- Do telecommunications contracts call for any "virtual private network" services such as:
  - AT&T's Software Defined Network (SDN) or VTNS
  - MCI Communications Corp.'s Vnet
  - Sprint Corp.'s VPN Premiere

  A "virtual private network" (VPN) or "software defined network" uses portions of the publicly switched telephone network in concert with dedicated private facilities. VPN subscribers are provided with the capabilities of an advanced private-line-based corporate network without the need for point-to-point transmission facilities over dedicated physical paths. The development of new types of switches and routing programs has led to a proliferation of VPNs during the past decade. Unfortunately, the term "virtual private network" or "VPN" is also used to describe a packet-based virtual network; these two uses are not related. VPN, in the older sense (e.g., AT&T's SDN), typically is set up entirely within the carrier's network without any particular modifications to premises equipment.
- To what extent, if any, do long-distance rates vary according to the distance between the origination and destination of the call?
- Do the charges paid for inbound 800 service vary according to the time duration of the individual call? If so, by what increment of time? Six seconds? 18 seconds? 60 seconds? Some longer interval? What about charges for outbound, direct-dialed long-distance calls?
- What was the firm's total annual voice (not data) long-distance expense, after contract discounts, last year?
- What was the total amount of Federal Excise Tax charged for long-distance minutes for a representative month? For example, AT&T's Billing Edge CD provides a convenient tax report that provides the total excise taxes paid for a specific month. The telecom manager or director should be retaining CDs from the carriers for at least 36 months.

If the annual amount spent on long-distance voice (minute based) is in the $3 million+ range, it might make sense to pursue a refund. Whether it is worthwhile at the lower levels depends on the level of record keeping. If an organization is highly distributed and has no centralized records, the effort to gather the information may exceed the value. This is another justification for the use of global, enterprisewide contracts.

How the Refund Works in Practice

Assuming that an organization has made sufficiently large telecom federal excise payments to justify the filing, the process works roughly as follows:

- The Federal Excise Tax payments over a 36-month filing period are totaled to determine the total potential refund.
- If data for just a few periods is missing, sometimes the IRS will accept interpolation, particularly if the monthly payments are reasonably consistent.

Exhibit 1.   Example Potential FET Savings Chart

Net annual voice long-distance payments to carriers (LD, calling cards, 800 number)	$7,000,000
Federal excise tax at 3 percent of total	$210,000
36 months of FET	$630,000
IRS settlement at Appellate at 39 percent	$245,700
Typical tax consulting (third party) contingent fee (33 percent)	$81,081
Net cash to organization for 36 months FET	$164,619

- Third-party tax consultants are usually selected by the organization based on expertise in this area (i.e., those who thoroughly understand the filing procedures and have the requisite *savoir faire* with the IRS).
- The tax consultants handle filing and appeals, typically with little or no additional assistance from the organization.
- Within 10 to 18 months, depending on the location of the filing, the IRS refunds between 25 and 39 percent of the Federal Excise Tax originally paid.

There have been real success stories in FET refunds. In some cases, the IRS has settled for up to 40 percent of the total FET paid to carriers. Exhibit 1 illustrates the potential value in pursuing this opportunity.

Nonprofit Organizations

Nonprofit organizations may be exempt from telecommunications taxes (as well as some other taxes), based on 501C classification. Service providers may inadvertently include taxes where none are due, so a review of taxes paid is essential for any nonprofit with substantial telecommunications expenses.

Exempt Businesses

Certain businesses are exempt from the Federal Excise Tax and should receive a full refund if they have been charged. These include:

- News services (general press, radio or television broadcasting)
- Common carriers (airlines, public trucking lines, etc.)
- Certain military operations

Appendix C, taken from the relevant IRS publication, outlines both nonprofit and for-profit exemption rules.

Summary

Telecom taxation and quantum mechanics have at least one thing in common — not many people understand the details. As a result, it is likely that if an organization is improperly billed for taxes, the error will remain undiscovered. By looking at the potential for refunds, exemptions, and the accuracy of the calculations themselves, significant cost savings can be identified.

Notes

1. This is one case where outsourcing is by far the most practical approach. Conquering the tax learning curve internally for some complex filings is simply not worth the managerial effort.

Chapter 9

RFPs, Contract Optimization, and Outsourcing

Honest disagreement is often a good sign of progress.

— Mahatma Gandhi

Written above the ivy-covered archways of many prestigious law schools is one of the basic legal truths: "Oral contracts are not worth the paper they are written on." As in other business arrangements, the details of the contract should be in writing rather than in the minds of the individuals at the negotiating table.[1] The sections below and Appendices E and F at the end of the book, outline key information elements that should be reviewed and made final before a contract is signed or an RFP (Request for Proposal) is released.

The rewards of spending a little time in the legal muck of contracts and telecom minutiae can be significant. Conducting a periodic review of in-place service and equipment provider contracts saves money — sometimes as much as 20 to 40 percent if the agreements have not been reviewed for several years. Some review areas include:

- Reevaluation of annual spend commitments
- Modification of existing contracts (rates, volumes, service levels)
- Evaluation of the benefit provided by a RFQ (Request for Quotation) or RFP

Dedicated and switched rates have decreased by half over the past few years; unfortunately, many firms fail to pursue the cheaper rates now available. Exhibit 1 shows potential savings from competitive bidding and contract negotiations.

Exhibit 1.   Possible Savings on Services Provided by Carriers

Equipment, Facilities, and Network Management Options	Estimated Possible Savings
Network equipment assets and options	4–25 percent
Emerging alternative technology options	10–60 percent
Service migration strategies	10–45 percent
Reviewing total cost of ownership strategies for staffing and management	4–20 percent

Throwing It over the Fence: Outsourcing

Outsourcing continues to prove popular with firms looking to assuage their worries by subrogating operations for voice and data communications, IT functions, help desk, etc. Like strong armies that prepare for war and thereby avoid it, the current state analysis required to enter into an outsourcing agreement could eliminate the need to outsource. For example, an "as-is" network study could initiate changes such as the consolidation of network management, centralized computing, and rationalization of the network. Hence, the study itself will likely achieve operational gains and save money whether or not the actual outsourcing takes place.

Whether the decision is made to outsource or to improve the efficiency of the in-house function, agreements and processes need to be reviewed. The following section lists some fruitful areas of review.

The Telecom Procurement/Contract Review

The following are guidelines and concepts that will likely provide many benefits if they are included in contracts. This is not an all-inclusive list of contract terms and conditions to review but serves as a sampler of some important contract elements:

- *Tiered pricing.* As the volume of purchases goes up, the discount applied off tariff should be greater. For example, domestic long-distance minutes might be 40 percent off tariff for the first 15 million minutes, 44 percent off for the next five million, and 46 percent off the next ten million minutes. Volume should always drive price.
- *Most favored customer clause.* The customer should get the lowest rates available to any of the vendor's customers of like volume and circumstances. If Billy Bob has 20 million minutes a year and gets 5.5 cents per minute on 800 service terminated to a dedicated location, then Mary Jane should get the same price, as long as she has the same volume of minutes and terminates in the same way (not switched).
- *Technology upgrade.* If the customer elects to use a newer technology, provided by the same carrier, to carry its traffic or perform other functions,

then there should not be a penalty for converting to the newer technology. For example, assume that a business is using T1s to carry voice traffic and then elects to use a new technology provided by the same carrier to carry the traffic (e.g., VoIP). In such situations, penalties for drops below minimum requirements should be waived.

- *Renegotiation.* There should be at least an 18-month annual renegotiation for a long-duration contract. The parties should negotiate in good faith to ensure that pricing is "competitive" in the marketplace. Many contracts specify 12 months and some firms, for example, have the right to renegotiate every six months.
- *Business downturn.* If a division or business unit is sold or discontinued, the minimum annual commitments should be reduced by a *pro-rata* amount.
- *Poison pill.* Avoid a "poison pill" of zero discounts at some very high level of minutes that the vendor assures the buyer will never be reached. The purpose of the poison pill is to ensure that resellers do not grab the contract and resell the discounted minutes at a higher rate. In one situation, a firm's usage exceeded its wildest expectations and got hit with high incremental prices. It helps the carrier but does nothing for the customer.
- *Exclusive contract.* While it might be an advantage for the customer to use only one vendor, it is rarely advantageous to contractually specify that one firm is the sole provider. In fact, in a large organization, it is virtually impossible to police the network and ensure that a "renegade" department manager or field operator does not cut a deal with the local telco.
- *Special requests.* If the customer wants network maps, escalation procedures, vendor personnel on site, or other special requirements, putting those requests in the contract is the best approach.
- *Discount-usage match.* The value of the contract depends largely on how well the discounts on particular services match the actual usage. For example, 12 cents a minute to London from Denver is of little value if the company does only a couple of hundred minutes a month to that location. Categories include interstate, intrastate (varies by state), inbound, outbound, 800, calling card, domestic, international, Frame Relay, T1, T3, OC3, E1, etc.
- *Installation waivers.* Carrier installation costs should be waived for T1 and T3 installs (they may require the circuit to be in place for a year)
- *Penalties.* Carriers should always agree to a *pro-rata* refund if a leased line is down. Additional penalties can also be negotiated as part of a service level agreement.
- *Minimum annual charges.* The customer should be reasonably certain that minimums will be reached to avoid penalties. Negotiate for lower minimums. Consider the possibility that dedicated circuits may be economically justified; if X number of minutes to a specific location (e.g., Paris, France) are committed in the contract, the loss of those minutes could result in a penalty.
- *Sub-minimums.* Avoid excessive sub-minimums — some carriers require a specific quantity of 800-number minutes, switched minutes, Frame Relay circuit dollars, etc. The customer can get locked into a confusing hodgepodge of minimum commitments that must be monitored. Ideally, there should only be a few minimums or one large-dollar minimum (large-volume minimum).

- *Audio conferencing.* Consider carefully the IXC's audio conferencing service. If it is on par with other external firms, then it may be beneficial to add those minutes into the contract. However, if a large volume of minutes is used, the client may want to consider using in-house audio conferencing where the incremental "meet me" bridge cost is zero (of course, the up-front equipment investment as well as administration time must also be considered).

- *Ramp-up period.* If a customer has multiple carriers that are being consolidated into a single carrier (usually the most economical alternative), the contract should specify a reasonable "ramp-up" period. During this time, the customer can convert existing agreements to the new contract and identify all relevant locations (more difficult than it appears). Pricing during the ramp-up period should be no different than when all volume commitments have been met.

- *Preparation for contract negotiations.* The more information on volumes (particularly international locations), the better the deal a carrier can offer (if they feel the competitive pressure). Volumes (minutes) should be available as follows:
  - Interstate
  - Intrastate (by state)
  - International (by country)
  - Switched, dedicated, and "mixed" traffic
  - Audio conferencing
  - Inbound
  - Outbound
  - Toll-free volumes (domestic and international)
  - Directory assistance
  - Cellular long-distance
  - Video
  - Calling card (also by categories)
  - Data circuits: T1, T3, OC3, Frame Relay, ATM, etc.

- *Ancillary services.* These services should be defined and agreed upon. For example: Who issues the calling cards? How do calling cards get billed back to the individual business units/employees? Who deals with urgent matters (e.g., it appears that a card has been stolen — who authorizes cancellation of the card and issues a new card)? Who works out the procedures to cancel cards when employees terminate?

- *Toll fraud monitoring.* Does the carrier monitor for toll fraud? Is there a list of key employees at every relevant location that can make a decision on what facilities to keep open or shut down if toll fraud is occurring?

- *Toll fraud insurance.* The carrier should provide toll fraud insurance. Deductibles should not be excessive (e.g., not more than $15K to $20K per incident). Review the terms to ensure that the organization can comply and not have a false sense of security. For example, most toll fraud insurance terms require that DISA be disabled.

- *Combined services.* If the carrier offers both IXC and LEC services, there should be a significant reduction in pricing for those locations that elect to combine both services.

- *Service provisioning.* The carrier should maintain detailed electronic records of all orders (circuits, bandwidth required, locations, owner, characteristics of the circuit, etc.). Many are now offering browser-based packages that the customer can use to monitor the progress of the installation. The customer should receive regular status reports. No circuits should be implemented or disconnected without going through appropriate customer notification (change control).

- *Network optimization.* The carrier should commit to a periodic (quarterly; semiannual or at least annual) optimization review. For example, are there two T1s that are going from HQ to the same city but owned by different business units? Could they share the T1? Are there switched locations that can be converted to dedicated locations (this is critical and should be an ongoing review process)?
- *Reporting.* The carrier should provide extensive monthly reports showing volumes, commitment compliance, trends, and any management issues.
- *Calling cards.* Plans should be examined for options such as an 800 number to get into the carrier's network. With this feature, setup costs should be significantly reduced. A better option (if the carrier's billing system can do it) is to have the customer employee dial 0+ and have the carrier's network recognize that it is a card on "XYZ's" corporate plan and automatically provide the lower setup fee.
- *Billing details.* Is billing in six-second increments? Is there a minimum of 18 seconds? Does the firm have applications (modems) that have minimal duration calls?
- *Options.* What financial options are available? Are there up-front credits? Is there a bonus when certain volumes are reached?
- *Exception reports.* Will the carrier run regular exception reports such as longest calls (maybe modems got "stuck")? Calls by area code or city? By type of traffic?
- *Carrier international relationships.* What international relationships does the carrier have? Does the carrier have any global plans for specific international cities?
- *Billing details.* What are the nitty-gritty billing rules? For example, if a fax machine tries multiple times to reach a location (most relevant for international faxes), does the carrier bill for repeated tries or only for actual minutes after connection?
- *Single points of contact.* Will the carrier identify an individual to be the customer contact point for troubleshooting?
- *Rollover terms.* What are the rollover terms? Does the contract stop on the termination date or roll over if no notification within 90 days?
- *Mobile phone negotiations.* Will the cellular provider PICC[2] all long-distance calls to the organization's carrier? What are the time of day/weekend terms? What are the roaming conditions? Is there an on-site service rep? Can executives get premium support? Does equipment always have to be ordered from an out-of-town location, or is there a store on hand for emergencies/executive needs? Does the cellular provider have GSM phones? Are the GSM phones linked to the employee's account? How is billing done — individual statement or mass bill? How are defaults handled (employees who leave the firm or make phone calls they cannot pay for — yet the company has "guaranteed" payment of the bill to get the lowest rates)?

The Request for Proposal/Request for Quotation

The individual responsible for initiating an RFP/RFQ should consider the following "philosophy" questions before moving forward:

- *What is the end result?* Lower costs, better service, specific functional requirement, or perhaps (legitimately) to demonstrate that the current provider is the best and no others can compete.
- *Is it worth it?* RFPs/RFQs are time-consuming for all parties. Sometimes it is better just to buy a hammer at the nearest store, regardless of its price. What economic or "political" conditions would make an RFP/RFQ mandatory?
- *Is the document to be highly detailed?* Is there enough information to write a worthwhile RFP, or will vendors be spending fruitless hours guessing what is required? For example, when bidding on a large PBX installation, the vendor needs to understand the number of digital and analog lines required because these drive the quantity of line cards, shelves, etc. If the RFP is too high level, one of two outcomes will result: (1) some potentially strong vendors will not participate or (2) bidding vendors will increase their prices to cover unforeseen costs. A third, less-savory practice is the deliberate underbidding on a vague RFP with the expectation that profits will accrue from the many change orders that the vendor knows will be required.
- *How can a good list of suppliers be obtained?* Should a two-step process be followed, with a first wave of perhaps ten respondents and a second tier of three for the most thorough review?
- *How is the evaluation to be done?* Who has the time, qualifications, and objectivity to do it?
- *How much education is required from the respondents?* Does the organization need key evaluators to attend demos, talk to references, etc.?
- *What can be done to make an admittedly "fuzzy" evaluation process more fair?* Will the "non-winners" be provided with a full explanation of the final decision?
- *Will proposal content from the winning bidder go directly into the contract?* This is an important requirement for the buyer because it ensures that the features and benefits presented in the proposal become contractual requirements.
- *Will price increases be allowed?* If the bidding and evaluation process takes several months, it is possible that the respondents' costs have increased in the meantime.

Creating the RFP/RFQ

Many firms have their own general terms and conditions such as insurance coverage requirements, confidentiality, and indemnification clauses available from their procurement department. In certain cases, boilerplate text is available. As expected, service providers have their own contract terms and conditions and some may insist on customer acceptance "as is." Smaller firms have less leverage and often are presented with the "take it or leave it" alternative. However, in many cases, buyers can negotiate changes.

The RFP process can be used for requesting innovative solutions in architecture, local and long-distance services, and new product deployment. Given the rate of technological innovation and carrier consolidations and divestitures, customers may not be aware of new products and prices available. Serving as the Swiss army knife of buyers, the RFQ is used for diverse purchases, such as installation, maintenance, equipment, and voicemail systems.

There has not always been a fine distinction between RFP and RFQ. In theory, an RFP invites the service provider to develop imaginative solutions to the buyer's business needs. In contrast, the RFQ states exactly what the buyer requires (very detailed specifications) and is looking solely for pricing. Most RFPs and RFQs fall between these two ideal structures. Generally, if the buyer does not have the expertise of staff to develop a well-defined RFQ, then an RFP is preferable. It does not make sense to send out an RFQ for yesterday's solution; reputable vendors know the "new, new thing" on the market.

Whether documents are simple or formal, they should contain explicit instructions for the respondent's proposed solution. Otherwise, the vendors are guessing requirements and the playing field for the subsequent evaluation is uneven. As telecommunications consultants, the authors have occasionally been brought in after an RFP has been issued to discuss the selection process with the "non-winners." Where bad feelings occur, the cause is most often poor wording of requirements and concerns that the selection criteria were vague, rendering the final choice as arbitrary.

Evaluating Vendor Proposals

How can one be fair? One choice, albeit not a particularly good one, is to assign responsibility solely to the purchasing department. The obvious problem with this approach is the lack of telecommunications expertise. A better solution is to develop a team with business, telecom, and other specialty knowledge to make the decisions. Ideally, these individuals will have previous experience obtaining and documenting the requirements for a distributed RFP/RFQ. They would then be the logical candidates as members of an evaluation committee for scoring the vendor responses.

Objective ranking is rarely easy. The potential for favoritism is always a concern to vendors. To increase at least the perception of objectivity, some firms aggregate individual evaluations of team members. One of the difficulties with any evaluation that purports to be completely objective is that some important factors, such as reliability and fit with organizational chemistry, cannot be quantified.

Scoring Model for Vendor Proposals

The previous paragraph not withstanding, a scoring model should be developed to ensure that all parties know what is important. For public organizations (government and nonprofits), evaluation procedures are *de rigueur*. Exhibit 2 illustrates an example score sheet for a telephone equipment RFP.

Exhibit 2.   Sample RFP Equipment Score Sheet

Category	Score	Weight (%)	Comments
Technical			
Hardware reliability and uptime	7	25	Exceeds the 99.999 percent requirement
Software quality	5	10	Industry reputation and experience of references checked suggests adequate quality
Compliance with standards (VPIM, QSIG,[a] etc.)	2.5	5	Supports VPIM but not QSIG; may not fully integrate with existing PBX systems
Availability of third-party hardware/ software for call center applications	10	10	All relevant applications, such as Blue Pumpkin for our call center, will run on the platform
Station features	5	20	Station features will support 99 percent of users; call park is not available
Call detail	10	5	Interfaces with all major call accounting/call management packages
Remote (wireless) services	2.5	5	The remote wireless offering is expensive and proprietary
Voicemail features	5	10	Middle of the road for large systems; unified messaging is a separate upgrade
Security (PBX and voicemail)	10	10	Strong security; interfaces with PBX firewall; password, monitoring, and auditing functions well developed
Weighted average for technical attributes	6.5	100	
Management and Background			
Market share	2.5	10	5th in market share for U.S.
Experience in call centers	7.5	25	Strength, particularly in small call centers with heavy CTI applications
Financial stability, going concern	6	15	Middle of the pack in terms of financial stability, cash on hand, and Standard & Poors evaluations, etc.
Knowledge of industry	6	15	Has worked in energy sector for several years
Understanding of requirements	9	20	Demonstrated thorough understanding of RFP and asked questions suggesting considerable research into the needs of the company

Exhibit 2. Sample RFP Equipment Score Sheet (Continued)

Category	Score	Weight (%)	Comments
Qualifications of implementation/ support team	6	15	Team appears to have adequate background in large PBX installations
Weighted average for management and background	6.6	100	
Support			
Logistics, including proximity to field offices and parts availability	5	35	Offices are distributed throughout the U.S. and are close enough to our plants to provide a reasonably fast response to problems
Service levels	7.5	45	Good response times and escalation procedures
Implementation schedule	2.5	20	Will require four months to begin installation
Weighted average for support	5.6	100	
Pricing			
Equipment initial cost	5	50	Equipment is only modestly discounted
Ongoing maintenance	10	30	Strong offering for 3- to 5-year maintenance agreement
Upgrades and expansion of initial system	7.5	1	Good discount level for future purchases
Technology upgrade	10	10	Excellent plan to protect company from technical obsolescence
Weighted average for pricing	7.3	100	
Weighting for Major Categories			
Technical		25	
Management and background		20	
Support		15	
Pricing		40	
Total		100	

Grand Total Score for Respondent #1 6.69

a The QSIG protocol provides signaling for private integrated services network exchange (PINX) devices. Among other things, it provides transparent support for supplementary PBX services so that proprietary PBX features are not lost when connecting PBXs to different manufacturers' equipment.

Exhibit 2 uses weighted average to ensure that the final score reflects the importance of one attribute over another. The individual categories have a sub-rating (e.g., 6.5 for the technical response) and then a weighting for all the sub-ratings (e.g., of the four major categories, pricing accounts for

40 percent of the "influence"). At the end of the table, a grand ranking of 6.69 is calculated for Respondent #1. If there are a total of five respondents, then five rankings will be calculated. Theoretically, the respondent with the highest ranking will be awarded the contract.

Outsourcing

Outsourcing options range from simple call accounting to complete responsibility for telecommunications (staffing, provisioning, negotiating with carriers, and efficiency studies). QuantumShift, for example, offers to become the telecommunications function of its customers. By negotiating with carriers, understanding the details of cost savings, and managing assets, the high-end service provider can presumably offer a better total package than end customers can obtain for themselves. All the usual pros and cons of outsourcing apply:

- Pros:
  - The provider is specialized in telecommunications and cost management, and thus develops expertise.
  - Costs are controlled by leveraging relationships with other providers (e.g., AT&T, MCI, Avaya).
  - Outsource staff have a career path in their specialty, whereas within the client organization telecom staff have limited upward movement opportunities.
  - The provider can leverage processes learned across multiple clients.
  - The provider may make specialized software and hardware available so that the client avoids some up-front capital expense.
  - Reporting, cost distribution, and call accounting are well developed and usually provided via Web browser screens.
- Cons:
  - The interests of the customer and the provider may not exactly match.
  - There is some loss of control over the process.
  - Technology enhancements could take a back seat to well-defined, transaction-oriented charges. In other words, if the entire contract is written so that the service provider is paid solely on a defined transaction basis (interstate minutes, number of T1s, etc.), there could be a tendency to maintain the status quo. Codicils should be put in place that maintain incentives to continually review new technology and implement as appropriate.

The decision to outsource will vary by company and circumstances. Small to mid-sized firms, in particular, may find it convenient and cost-effective to outsource all or parts of their telecommunications functions. But, as they said in ancient Rome — *caveat emptor.*[3]

Summary

Substantial savings and quality improvements accrue to those organizations that take the time to ask for what they need, evaluate the responses, and monitor the results. Although the purchaser of telecommunications services and equipment has many challenges — ambiguous or complex feature sets, unknown quantities, and uncertainties about technology direction — a strong RFP can provide the structure for getting close to the optimum deal.

Outsourcing is a purchase as well, but on a larger scale. Using a structured approach will allow the organization to buy the quantity and quality of resources needed.

Notes

1. Of course we mention here the expected caveat — the authors are not attorneys and the reader should obtain legal advice before entering into an agreement for services with any telecom vendor.
2. Primary interexchange carrier. Cellular telephone companies will send long-distance traffic to the customer's preferred long-distance carrier if it has been set up. Otherwise, the long-distance call will be charged at the cellular telephone company's full retail rates. Lately this issue has become less important as cellular firms are increasingly bundling long-distance with minutes in the nationwide, flat-rate plans.
3. Let the buyer beware.

Chapter 10

The Future

Big brother is watching you.

— George Orwell, *Nineteen Eighty-Four*

Professional futurists — at least the honest ones — usually hedge their divinations with caveats and supply relatively vague notions of where the telecom industry is headed. Our Ouija board is no better than theirs. What we can do is look at some trends that will most certainly affect costs and feature availability in the next three to five years. We know that some basic laws, such as "there is no free lunch," will not be repealed, contrary to the expectations of the "free calls over the Internet" enthusiasts.[1] Other principles, such as "consolidation wins" for providing the cheapest telecommunications services, may or may not be repealed before the end of this decade.

Following are structural changes and trends we anticipate over the near to medium term. Some are obvious; some may be counterintuitive; some will undoubtedly prove false. We welcome reader comments.

Regulatory Changes

The U.S. Telecommunications Act of 1996 was the death knell of the "monopoly is a good thing for telecom" philosophy. Many new rules were introduced to jump-start competition. Other nations are following suit at varying rates. Monopoly is out and deregulation is in. Some of the expected results over the next few years include:

- *Regulatory convergence.* Too many regulating bodies confuse and slow the intended growth of free markets. The United States has long been hampered by a struggle between state regulations, the FCC, and the courts. Expect a painful but slowly accelerating rationalization of regulation, perhaps leading to more effective, rather than merely theoretical, competition at the local level.

- *Relaxation of barriers to entry.* The regulators have had the paradoxical task of introducing more regulations (such as forcing the ILECs to wholesale services to CLECs) in order to ultimately increase competition so that fewer regulations are required in the long run. Depending on the extent to which this strategy works, more telecommunications providers can enter the market, resulting in lower prices.

- *Relaxation of local service regulations.* The history of cellular service provides insight into this potential trend. The FCC licensed the players and then kicked them out of the nest — no rate controls, unbundling requirements, or mandated resale. The result was lower prices and even competition for local wire line services. The case for local regulatory supervision is being slowly eroded.

- *Universal service offerings (USO) will become broader.* Traditionally, USO has equated to the practice of slightly overcharging in the cities to subsidize simple telephone service in the countryside. As other services, such as broadband Internet access, become *de rigueur* for businesses and even individuals, some expansion of the urban tax is likely. While this may slightly increase taxes paid by many corporations, it also provides more opportunities to relocate in rural areas that may have lower labor rates, since because essential communications services will be available. The downside to this "Robin Hood" approach is that the calculation of the appropriate amount of the tax is difficult.

A World Lit Only by Packets

In William Manchester's book *A World Lit Only by Fire,* he talks about the extreme isolation of the medieval world, where many villagers never moved more than a ten-mile radius from where they were born. In some ways, our circuit-switched world has been equally confining. With the old point-to-point orientation, communications capacity and cost were linearly related. More bandwidth could only be obtained by the brute force of more lines or channels. The poorest people could not easily be reached because it literally cost more than their "net worth" to connect to them.

With packetization of communications, the world changes in fundamental ways. Now, many channels can be shared and any one channel can carry packets from a myriad of sources and media types. Some of the implications of this fundamental change include:

- Given the at least 10:1 cost advantage of IP^2 transport over circuit-switched (TDM), unit costs will continue to fall. Much of the world's communications are still based on circuit-switched technology, so there is a long way to go.

- Any-to-any communications will grow because it is easier to convert packets from one media to another. For example, an increasing number of the world's faxes never see paper — they are captured electronically; e-mails are converted to voice via text-to-voice. Video, once a luxury only justified by reduced air travel, is rapidly becoming an IP commodity. Machines, storage devices, and humans will all network in increasingly complex ways.

■ Many more devices will communicate. New protocols and technologies, all dependent on packetization, will greatly expand the network. For example, the Bluetooth specification allows handheld devices like Palm Pilots to communicate with like-enabled telephones; when the owner walks into the office, the devices talk, sharing information such as new directory entries, missed telephone calls, etc.

Keynes Was Right

In 1923, John Maynard Keynes, the noted economist, stated that the "long run is a misleading guide to current affairs. In the long run, we are all dead." The underlying principle of this statement — that the long run is too unpredictable to factor into current calculations — applies to telecommunications as well.

In the past, PBXs were amortized for ten years without fear of obsolescence. Telecom directors, thinking they were doing the right thing by buying top-quality stuff that would last a long time, paid a premium for proprietary equipment. The equipment was typically used, with little change, for most of the amortization period. Now, however, just as Keynes noted, there is no "long run." Sinking large dollars into equipment that could easily become obsolete is risky. Consider two CLECs: one CLEC buys a traditional switch such as an Avaya 5ESS-2000; the other purchases a much less-expensive "soft switch." While the soft switch, based on a more flexible technology, may not have the features or 99.999 percent reliability of the 5ESS,[3] it may be good enough for the "short run." And when new technologies that reduce costs are available, the CLEC that bought the soft switch may be able to chuck it in the Mississippi River and buy the newest technology. With a much higher up-front cost, switching infrastructures quickly may not be practical.

The net effect of the industry focus away from the long run is dramatic pressure on the hardware/software vendors to create products that can be amortized much more quickly and be more easily fitted to new technologies. All telecom buyers, whether providers or consumers, will look with a jaundiced eye on fixed telecom investments that must be made practical by long amortizations. Telecom now lives in the short run.

Storage Virtualization

Increasingly, organizations are turning to virtual storage. Although data is physically always stored somewhere, logically it can be considered as a cloud so that, for a fee, the end organization deals with just the logical data, not the file structures, host operating systems, etc. The reason this is important to telecommunications is that, increasingly, the physical data will reside outside the organization's premises. Communication links are essential to get to the data.

The following characteristics of virtual storage, courtesy of R. Baird of Hewlett-Packard Corporation, highlight some of its desirable features and why usage is likely to increase. Objectives include:

- *Separate usage from location.* Storage devices and media should be attached to the network and shared by all systems.
- *Separate logical from physical.* Storage objects such as databases and file systems should be distributable with no inherent physical boundaries limiting distribution.
- *Separate presentation from implementation.* Storage objects such as databases and file systems should be internally replicated and relocated as needed.
- *Separate policy from mechanism.* Storage objects should have minimally hard-wired behavior.
- *Separate control from data.* Data should move directly between source and sink (beginning point to ending point) locations, regardless of the number of intermediate mappings.

As storage moves towards "get me 100 gig more storage in the next five minutes" and away from "let's add another big hard drive to server HOU22 next week," requirements for both bandwidth and reliability of communications links will increase.

General Trends

Many of the general technology trends apply equally well to telecommunications and have a similar effect on costs. The following list includes the major shifts that should continue for the next three to six years:

- *Overall decreased regulation.*
- *Ubiquitous computing/telecommunications.* Failed efforts such as Iridium not withstanding,[4] the globe is moving slowly to communications anyplace, anytime.
- *Innovation at the hardware and software levels will continue unabated.* Compression techniques, improvement in fiber-optic protocols and endpoint equipment, and intelligence at the "edge" of the network (smarter routers and other data communications equipment) will ease the transition from a segregated voice/data/video world to a multimedia transport wonderland.
- *Packet-based telephony will spread consistently.*
- *Other forms of communication, such as instant messaging, will become more mainstream.* Commodity and other traders have used devices such as Amtel's direct line systems for several years (see Exhibit 1). When a trader has a phone in each ear, a pager vibrating, a co-worker yelling, etc. — sometimes instant messaging is the only way to say "Lunch — 11:30 — OK?" Japanese school children, often technical trendsetters, use NTT DoCoMo's i-mode cell phones to send text messages as well as play games and music.

Exhibit 1. Amtel's "Direct Line" For Instant Messaging

- *Contact of telecom providers as well as other businesses with the end customer will change.* Instead of a separate billing and service group, the customer will more frequently call a single point of contact (reducing time spent for both customer and provider). The use of automation, although not always appreciated by the customer, will continue to increase. IVR (interactive voice response) systems and voice recognition systems will carry an even larger portion of the workload.
- *Outsourcing will continue.* Companies like Profitline, QuantumShift, and Stonehouse Technologies will manage telecom bill payment, procurement, ongoing cost analysis, negotiation with carriers, and even provisioning of equipment.
- *The rise of more application service providers will increase telecommunications uptime and bandwidth demands.* Web hosting, ad management, data analysis, and many other complex functions (or those with a high start-up cost) lend themselves to ASPs.
- *Organizations with international communications needs will develop carrier arrangements that use a global approach rather than optimizing individual national services.* There are several drivers for this trend: (1) by combining worldwide volumes, better enterprisewide pricing can be obtained; (2) a global approach improves standardization and predictability; and (3) outliers, both from a cost- and service-level perspective, become obvious and can be more easily be corrected. While there are negatives of global contracts and management, such as the potential to lose contact with local market conditions, the positives will likely prevail.

Summary

Our story ends here. There are more than 50,000 books per year published in the United States alone and ours must take its place among all the other business and technical tomes that vie for the attention of busy managers and executives. The one message we leave with our information-overloaded readers is this: telecommunications costs are manageable and can almost always be lowered significantly. With the continuing restructuring of the telecommunications service providers and the steady pace of technical innovation, the payoff for ongoing review of telecommunications has never been greater.

Notes

1. Our focus in this book has been on reliable, bet-your-business, works-every-day communications. For a hobbyist or occasional user, many telecommunications features (voice, chat, video) are free, if one's time investment is ignored. Ultimately, however, people work for incentives and "free" must always equate to either hobby/public service interests or a form of advertising.

2. Internet Protocol. These are the specifications upon which the Internet was built.

3. The intent of this discussion is to show the effect of large, up-front investments in telecom capital equipment versus more flexible investments in short-term systems. Comparing a traditional CO type machine such as the 5ESS with a soft switch is complex and many issues are involved (scalability, reliability, features, etc.). The less-expensive option is not always the right answer. Also, equipment vendors supplying "traditional" telecom devices are finding ways to make them more adaptable to newer standards.

4. Iridium, a satellite-based telephone service created by Motorola, most likely failed because there was no real sustainable need for the business service/cost model. The technology itself, communications using low earth orbit satellites, is sound.

Appendix A

Telecommunications Cost Management Checklist

Toll-Free Numbers:

- Consider a VPN (virtual private network) solution for remote access versus a 1-800 dialup plan.
- Block local remote users from using the 1-800 dial-up number. Users should be given a local number to use instead.
- Terminate unused toll free numbers (avoid the fixed monthly fees).
- Consider terminating *intra*state 800 number traffic in another state and then backhauling across state lines to the physical location receiving the calls. This achieves savings by using less-expensive interstate instead of intrastate rates. Note that this savings is only applicable where a substantial portion of the traffic is intrastate.
- Review 800 numbers to ensure that they are terminated via dedicated access lines (e.g., T1) rather than "switched" access. In the case of AT&T, for example, this would be the difference between "Readyline" and "Megacom" service. In some cases volumes of traffic may not justify dedicated line termination.
- Review toll-free numbers requested to be removed and verify that they have actually been dropped from invoices.

Long Distance (IXC):

- Block premium numbers such as 900, 976, and 976 look-alike numbers to avoid potential toll fraud and employee abuse.
- Provide written notification to the local access carrier that any changes to long-distance service need to be approved in writing by an authorized individual(s) within the organization. Otherwise, the potential for slamming/cramming by other, sometimes high-cost long-distance carriers exists.

- Review usage of circuits to ensure they are continuing to be used (i.e., not going to a closed office).
- Ensure local billing has only local calls separate from third-party carriers.
- Review contracts (preferably in the negotiation phase) for incremental billing. For example, what is the minimum number of seconds billed?
- Determine if exception reports identify excessive duration calls. For example, calls over 12 hours could be the result of toll fraud, a "stuck" modem, or an employee who never shuts down an online remote connection ("I've got a good connection now and I'm not going to lose it!").

Local Service and PBX Options:

- Ensure that PBXs are set up with least cost routing. For example, assume that an organization has a dedicated "tie line" from its headquarters office in New York to a field office in Houston. If an employee dials the full ten-digit number rather than a shortened speed dial, the PBX will recognize where the call is going and route it automatically to the tie line rather than sending it through the public network.
- Review invoices from the local exchange carrier (LEC). If trunking to the long-distance carrier is inadequate to handle the traffic, most PBXs will complete the call by routing it over local access lines. This rerouted traffic is likely to be more expensive because the per-minute charges are at standard "mom and pop" rates from the LEC.
- Inventory and review POTS (plain old telephone service) line charges. These lines do not go through an organization's PBX, but instead get dial tone from the LEC's PBX (central office). While useful for emergency phones and some other purposes, they are expensive relative to analog lines provided by the organization's PBX and should not be used indiscriminately.

Trunking/Circuits:

- Evaluate the number of T1 trunks and consolidate into one or more T3 circuits if there are more than eight to ten T1s.
- Change out PBX cards to accept a PRI T1 rather than the older style T1. Typically, PRIs are less expensive and provide caller ID information.
- Consider using DSL as a Frame Relay access line rather than a T1. This will lower the access costs.

Cellular and Wireless:

- Establish enterprisewide agreements (volume discounts). In some cases, regional pricing may be the least costly option.
- Analyze utilization by cell phone; if many employees are significantly under- or over-utilizing contracted minutes, consider using a minute sharing agreement with the provider.
- Determine whether activation charges are waived for reused cell phones. If so, negotiate reduced or eliminated charges when a cell phone is used for another employee.

- Review cellular bills for excessive roaming charges and appropriate plans. Often, the highest charges are from employees with legitimate business reasons for making the calls but using the wrong volume plan.
- Determine if very high minute users can have two plans on the same cellular phone.
- Determine if an internal resource should have at least part of his or her time devoted to analysis of cellular expenses. For large organizations, this can have a substantial payoff.

Calling Cards:

- Review rates and setup charges.
- Set up 800 numbers so that employees and other authorized individuals can call the office or plant without using a calling card. Toll-free numbers have no setup charge and are usually cheaper per minute.
- Examine processes for the return of cards when an employee or other user leaves the organization.

Review telecom contracts for the following:

1. There should be a substantial cost difference between switched, mixed, and dedicated-to-dedicated traffic.
2. The contract should allow for any conceivable level of volume growth. For example, some carriers will argue for a cap on discounts above a certain volume of minutes (often far beyond the customer's expectations). The reason for this is that the carrier may not want to have a deeply discounted tariff that can be used by resellers, who will generate hundreds of millions of minutes in business.
3. Business downturns and unexpected business upswings should be addressed. There should be no retroactive penalties if, due to a business divestiture, anticipated volumes are not met.
4. The value of the contract depends largely on how well the discounts on particular services match the actual usage. For example, 10 cents a minute to London from Denver is of little value if the company does only a couple of hundred minutes a month to that location.
5. Carrier installation costs should be waived for T1, T3 installs (they may require the circuit to be in place for a year).
6. Service level agreements should be in place.
7. Carriers should always agree to a *pro rata* refund if a leased line is down. Additional penalties can also be negotiated as part of a service level agreement.
8. The organization should be reasonably certain that minimums will be reached to avoid penalties. Negotiate for lower minimums. Consider the possibility that dedicated circuits may be economically justified. If X number of minutes to a specific location (e.g., Paris, France) are committed in the contract, the loss of those minutes could result in a penalty.

9. Review volumes (minutes) for the following traffic types and ensure that the overall economic package is properly weighted based on volumes:
   a. Interstate
   b. Intrastate (by state)
   c. International (by country)
   d. Switched, dedicated, and "mixed" traffic
   e. Audio conferencing
   f. Inbound
   g. Outbound
   h. Toll-free volumes (domestic and international)
   i. Directory assistance
   j. Cellular long-distance
   k. Video
   l. Calling card (also by categories)
   m. Data circuits: T1, T3, OC3, Frame Relay, ATM, etc.
10. Ask for a "most favored customer" clause. This clause ensures that the customer gets the benefit of any price break or other consideration afforded any other carrier customer with similar volumes/circumstances.
11. Include a renegotiation clause. Ideally, customer and supplier will meet annually to adjust pricing based on market conditions or other changes in circumstances.
12. Review minimums and sub-minimums carefully. Some carriers may require a percentage of traffic to fall into certain categories. For example, 65 percent of all voice traffic (over the PSTN) may need to be "dedicated" rather than switched; or 50 percent of the volumes must be in toll free, international, and VPN. These numbers should be carefully reviewed to ensure they are reasonable in terms of anticipated growth, acquisitions, divestitures, and potential technology changes (such as a VoIP implementation).
13. Examine termination requirements. For example, five-year contracts are generally considered too long due to the decreasing unit costs of telecom services.
14. Perform volume trend analysis monthly. Make sure that minimums are met (if consistently not met, consider renegotiating the contract). Also review for large increases that may justify additional discounts for larger commitments.
15. Include on-site personnel to aid large-scale implementations, projects, or billing problems due to a change in business or a volume change. Such add-ons are increasingly difficult to obtain but certainly should be considered for large telecom users.
16. Ask for toll fraud insurance. In some cases (for large customers), the carriers can provide it for little cost.
17. Include a technology upgrade clause. This ensures that penalties will not be incurred if the customer upgrades services (technically) within the same carrier.
18. Consider obtaining IXC and LEC services from the same carrier. The result will sometimes be a reduced total cost, particularly for a large organization with many locations. Appropriate backup capability should be considered in the decision.
19. Tie contract terms and pricing to specific service level agreements.

20. Include waiver of access coordination fees for circuit installs.
21. Consider financial options such as up-front credits for consummating agreements. For organizations that are temporarily cash strapped, this option provides flexibility.
22. Investigate billing details. A few international carriers may charge for attempted connection rather than actual. For example, international fax services could charge for repeated attempts to destinations that have poor-quality lines
23. Review rollover terms to ensure that, at expiration, rates do not jump up drastically if sufficient notice is not provided to the carrier.

Frame Relay:

- Consider the use of asymmetric PVCs if traffic between locations is not equal (e.g., most traffic flows from HQ to field office or vice versa).
- Consider a combination of IXC and LEC Frame Relay to reduce costs. For example, HQ and the primary regional hubs could all be linked via the IXC's Frame Relay and the regional hubs to field offices could be linked via the cheaper LEC Frame Relay service. The downside is that some network management information may not be available when the network traverses two carriers.
- Negotiate with the carrier to pick up the per-minute charges for the ISDN backup during any Frame Relay downtime.
- Consider using a zero CIR PVC for a backup circuit.
- Examine the potential for replacement by MPLS (multiprotocol label switching) technology.

IP Telephony and VoIP:

- Review the pros and cons of installing an IP-based PBX in small- to medium-sized offices. Consider equipment costs, maintenance, and reduced expenditures as a result of combining data and voice wiring.
- Review the voice traffic costs between office or plant locations. VoIP gateways are now less than $2000 per site. If WAN links already exist, the economics are significantly improved.
- Consider Voiceover-Frame Relay where a frame network already exists. International calling over Frame Relay is particularly cost effective versus the PSTN.

WAN Protocols:

- Review WAN protocols for excessive overhead. Using a more advanced protocol (e.g., Cisco's EIGRP versus RIP) reduces the need for additional bandwidth.

Call Centers:

- Consider the use of workforce management software, such as Blue Pumpkin or TCS to most efficiently use agents (strike the optimum balance between service to callers and agent utilization).

- Use automated tools to the fullest extent consistent with management objectives. IVR (interactive voice response) units handle routine inquiries with far less expense than human agents; other technologies, such as CTI (computer telephony integration), speed delivery of information, sales, and services to callers.
- Investigate the potential of using outside call center providers to handle overflow.
- Consider load balancing between call centers to reduce the total number of agents required to handle peak volumes.

Miscellaneous:

- Carefully consider key systems versus PBX technology. Key systems are reliable and considerably less expensive. Buy PBX technology only if there is a business need.
- Use voice recognition by IVR systems (e.g., Syntellect or Veritel) to reset passwords, thereby reducing helpdesk hours.
- Use a voice name dialer to reduce operator time or administrative assistant time. Examples include Parlance and Lucent large-capacity name dialer.
- Use voice recognition by an adjunct such as Veritel to provide "safe" DISA for legitimate callers who want to get dial tone from the office and dial out. This can be cheaper than using a calling card and is also more convenient for the user. Example: an executive in London calls her personal residence in Houston by dialing the Houston office Veritel server, then getting dial tone to make a local call.
- Consider using IP Centrex. MAC (move, add, change) is a significant cost of Centrex. If the LEC offers IP-based Centrex service, the MAC cost is reduced because the phone is hooked into the LAN network and can be moved with little manual administration.
- Standardize on a single vendor for fax machines. The transmission speed for fax machines is increased when like machines use a proprietary protocol rather than the slower group III. For example, a long fax between a Pitney Bowes fax machine and a Ricoh fax machine may take 20 to 30 percent longer than a Pitney-to-Pitney or Ricoh-to-Ricoh transmission.
- Hotels and some other organizations rarely use "answer supervision" in their PBXs. This means that the PBX does not actually know whether the outgoing call was completed. In practice, it is assumed that if the phone is still active after 20 to 30 seconds, then the call was completed and the caller will be charged. If the caller lets the telephone on the other side ring for a long time, she will be charged for the call even though it never took place. Individuals who travel frequently should be on alert for this defect in the billing system for some hotels, hospitals, and other organizations that charge guests for telephone services.

Miscellaneous Carrier Charges:

- Review invoices for equipment and services that are no longer applicable. For example, a "Personal CO Line" is rarely needed today.
- Consider a full-scope, detailed billing audit where the most detailed charges (termed USOCs for universal service order codes) are examined for applicability.

Appendix B

White Papers from QuantumShift

The following white papers, courtesy of QuantumShift (www.quantumshift.com), provide one vendor's perspective on telecommunications costs and the benefits of outsourcing telecom management. In the interests of space and format, some modifications have been made to the originals.

White Paper 1: What CEOs and CFOs Need to Know about Unmanaged Telecommunications Expenses

Executive Summary

In an increasingly Internet- and data-driven world, voice and data communications planning for the long term is essential. Yet it is often overlooked or given insufficient attention as companies grow. This is understandable given the explosion in technology and service choices. But this oversight can lead to serious business problems and uncontrolled expenses when communications services fail to meet a growing company's needs. Executives who are reshaping their firms to be more competitive want to put solutions in place that will support the company's growth and ensure its continued success. By outsourcing the management of telecommunications to an expert, they can:

- Avoid voice and data service-related problems that limit their growth
- Control expenses
- Free themselves to focus on their core business

The Management Challenge

Every company wants to cut costs and gain an edge on the competition. These bottom-line issues are the concern of executives who want to reshape their companies to meet the challenges of the new economy. In order to do this, they must get all parts of the organization focused and working together to achieve strategic objectives. But no company, regardless of its size or sophistication, can realistically hope to be the best in all facets of its operation, especially in today's complex and fast-paced business environment.[1] The challenge is to put the systems and processes in place that will enable a company's workforce to do the right things, not just more things, faster. Often that means strategically applying technology and outside resources to manage operations that are not a part of that company's core business.

Electronic commerce, e-mail, online customer support, telecommuters, and a mobile sales force are disruptive forces driving telecommunications needs today. New voice and data services are replacing old ways of holding meetings, sending messages, placing and fulfilling orders, and transacting business in general. As such, telecommunications is fast becoming one of the largest expenses and most complex areas of management for companies today. It is no longer just a matter of adding more telephones and trying to get the lowest rate on minutes. If the telecommunications infrastructure is not adequate to meet the escalating demand placed on it, it will be an inhibitor of a company's growth.

But the process of procuring and deploying telecommunications services is, alas, firmly rooted in the past despite becoming increasingly complex and frustrating, particularly for small and mid-sized companies. According to a July 2000 Forrester Research report, procuring telecommunications is "a pain for everyone." Companies of all sizes are unhappy at every stage of the procurement process, from ordering to installation to billing. Midsize companies in particular face difficulty because they lack the teams and the time to deal with technology complexities. The shortage of knowledgeable sales reps on the part of service provider worsens this situation.[2]

Effective purchasing and management of telecommunications is complex because it means:

- Researching service options
- Understanding and deciding which technologies to deploy
- Forging domestic and international carrier relationships for each service
- Overseeing timely deployment
- Ensuring remote access for telecommuters and the sales force
- Managing hundreds of contracts and invoices

The process quickly becomes expensive and diverts critical in-house resources away from core business needs. For many companies, it makes good sense to outsource the procurement, deployment, and management of communications services. Smart management in this case means finding the right partner who can assess a company's needs, provide the services, scale up, and maintain seamless reporting structures.

Outsourcing Non-core Functions

Outsourcing, once regarded as a strategy for merely saving money, is now viewed as an efficient way to get the right talent working on a task as well.[3] Tightly focused companies that employ outsourcing strategies are being rewarded, both with public capital and with market share.

The roots of outsourcing lie in out-tasking and subcontracting, but perhaps the concept is more fundamental: outsourcing can be thought of as a classic make-or-buy decision. As such, companies can assign real-dollar costs to both in-house efforts and outsourced resources. What is more difficult to quantify is the diversion of internal resources to non-core activities. Nevertheless, the costs are real and must be considered.

It is important to recognize that outsourcing is a long-term strategy, which differs from the transactional nature of subcontracting. Outsourcing is better modeled as an alliance of parties with shared interests in which each party incurs risk and shares the rewards. The success of this model rests on a company's choosing a partner that truly understands the business needs and has the insight, experience, and tools to provide a solution that will be viable in the immediate present and in the future.

Outsourcing is especially relevant to mid-sized companies that have many of the needs of a larger corporation but may not be able to support a dedicated staff. This is particularly true in the area of telecommunications management, where the tasks are urgent and complex, but the available pool of experienced talent is increasingly scarce and highly sought after.

The Breaking Point

Midsize and growing organizations run the risk of eclipsing the capabilities of the systems that got them to where they are today. Along with growth in business comes expansion into new geographies, and all this puts more demand on facilities, MIS, and telecommunications management. While experienced management teams see a "breaking point" ahead and plan system overhauls accordingly, often these plans call for scaling functions such as general ledger, human resource management, and sales force automation, but omit telecommunications management. This omission can be costly. Consider the following facts about voice and data services (see Exhibit 1):

- They are usually the second largest indirect expense within a company (third or fourth overall).
- They typically cost from $2400 to $4800 per employee per year.
- They are absolutely essential for day-to-day operations.

Detailed surveys done by AMI Partners of New York (see Exhibit 2) show that when telecommunications costs and complexity are compared to company growth, at various points the costs consistently overtake growth in companies of 400 to 500 employees. For example, Internet spending for companies of this size increases, on average, more than 140 percent.[4]

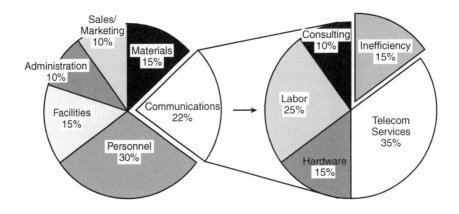

Exhibit 1.   Ineffectively Managed Communications Expenses

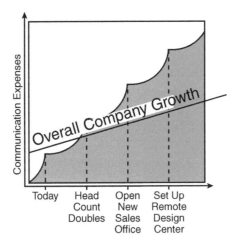

Exhibit 2.   Exploding Communications Expenses Due to Lack of Proactive Management

Without a strategic plan for scaling the communications infrastructure, sharply rising demand for services can lead to failure of an overburdened system, short-term losses and quick fixes, and longer-term escalation of costs.

Outsourced IT Set a Precedent

Not too long ago, IT functions were handled almost exclusively in-house, sometimes with the assistance of a patchwork of vendors. As the world became computer based and IT demands skyrocketed, many departments quickly adopted the ASP model, which recognized a shortage of IT expertise, the need to deploy applications faster, and the need to re-focus company resources on core competencies. Companies began to outsource much or all of their IT functions, which allowed them to gain expertise cost effectively in areas that brought high value to the company, but which were outside the realm of the company's core business.

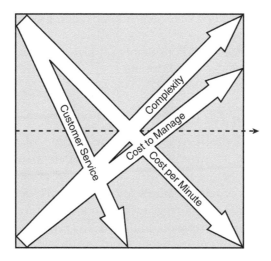

Exhibit 3. Total Cost of Service Increased by Decrease in Customer Support

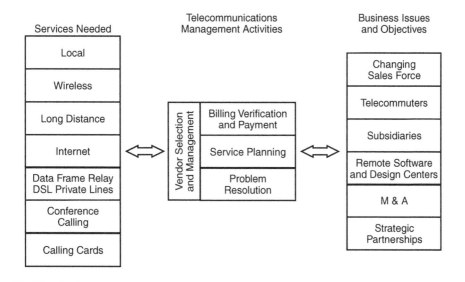

Exhibit 4. Communications' Effect on the Enterprise

In a similar way, voice and data communications have experienced exponential growth. But procuring and managing these services occurs in a tumultuous environment characterized by numerous vendors and widespread dissatisfaction among users (see Exhibit 3).

As Forrester Research points out, companies say that they are having serious problems obtaining the right services, negotiating and administering contracts, billing, and even getting basic sales and customer service — problems which are causing frustration and wasting time and money.

The incredible complexity of telecommunications management, illustrated in Exhibit 4, has led to a number of responses:

- The do-it-yourself approach, which keeps all telecom shopping and management in-house
- The use of auditors and consultants on a project basis
- The desire for "one-stop" shopping

The Missing Million Faces

A first response to the problem is to put the burden of telecommunications support on the MIS or IT manager. But with about a million IT jobs going unfilled in the United States today, the accelerated pace of business, and a rapidly shrinking pool of telecommunications talent, IT departments often just don't have the resources to manage another complex function. With many IT departments involved in high-profile initiatives to launch E-commerce ventures, the addition of telecommunications support can be a significant distraction.

A second response is to pull in telecommunications consultants to execute individual projects. This offers a solution to individual problems, but doesn't address the long-term needs for managing the infrastructure and services on an ongoing basis.

The desire for one-stop shopping is understandable, but companies must be careful to make informed decisions when choosing this approach. A single vendor may offer a range of communications services, but companies must be certain that these services are the right ones to meet fast-changing business needs over time.

A Communications Approach

With smart, proactive management of telecommunications services, companies can turn potential telecommunications woes into a tremendous edge. Successful management is built on three key requirements:

- Availability of professional services
- Careful vendor selection and management
- Modern, integrated management technology and tools

By combining technology, large-scale buying power, and the expertise of experienced professionals to simplify and automate the telecommunications management process, a strong telemanagement firm provides a single source for procuring best-of-breed voice (including wireless), data, and Internet communications. Service providers can be evaluated, selected, and managed. Ideally, the telemanagement firm will have a software platform that provides a comprehensive and click-simple management system. All services are integrated under one bill, enabling seamless planning, ordering, implementation, and expansion of services.

Other desirable attributes of a telemanagement service include:

- Telecom services for a distributed, geographically dispersed operation can be ordered and implemented quickly from a computer at a central location.

- A Web-based platform is available so that users can simply point and click to add, move, or change voice and data services.
- Detailed, online billing information is available, down to the individual user level.

Contracts: A Cautionary Tale

A corporation signed a three-year general contract with a major carrier with the intention of covering all the company's telecommunications needs, including those of the regional branches. But after the contract was signed, regional offices began negotiating on their own to obtain better deals on services — without informing the corporate manager responsible for the original contract. The regional offices did a good job negotiating, too, because they used the master corporate contract as a reference. Unfortunately, none of the negotiated regional contracts applied to the master contract.

The carrier is now holding the corporation responsible for each individual contract. And because the corporation has not met its commitments under the master contract for some time, it is in default and must pay a large amount just to stay in the contract or face renegotiating with the carrier. The carrier is trying to enforce the letter of the contract and the customer is understandably unhappy.

You might well ask, as this corporation did, why the carrier allowed such a mix-up to happen in the first place. Didn't the carrier's sales representative realize that the client was already a customer? Why didn't he bring the situation to light? So far, no satisfactory answer has been received.

The Bottom Line

So far, this story does not have a happy ending. The corporation is writing letters to the carrier, decrying the carrier's lack of shared responsibility and concern in neglecting to inform its client that the commitments of the original contract were not being met. Beyond that, the corporation has hired QuantumShift to manage their communications services.

Authors' note: One of the authors had a similar experience with a carrier but the misunderstanding was based on discount rates when volumes increased to unexpected levels. While an equitable arrangement was eventually worked out, the management effort and stress on all concerned illustrates the need for a thorough understanding of contracts and related issues.

White Paper 2: Managed Communications: Managing Communications for Greater Efficiency and Lower Costs

Executive Summary

In today's business environment, communications have become increasingly complex and costly to manage. Focusing on equipment and services, traditional providers of broadband, wireless, local, and long-distance voice and data services simply do not offer the solution companies need to effectively manage services. A new generation of communications providers is emerging, offering a single source solution that integrates the procurement, design/implementation, and management of communications services.

Part One: The Communications Challenge

Today, companies are leveraging the Web to restructure and automate every critical area of their business, from sales to finances to human resources. But there is one area of the enterprise that many companies are just now facing. Many experts believe that communications are the next major crisis that businesses will have to address.

New Demands for Communications

Services

Over the last decade, communications have grown from a few simple voice applications to enterprisewide mission-critical voice and data infrastructure. Remote access to the network, wireless interactions with customers, data connections between offices, a Web-portal for partners — every action your company takes is now part of a larger companywide, business communications game plan that integrates a wide spectrum of services: wireless, DSL, VPN, ATM, Frame Relay, PBX trunks, ISDN PRI, ISDN BRI, Centrex, voicemail, unified messaging, and more.

A few years ago these services weren't even among the top ten expenses that businesses worried about. Now, communications have suddenly become the company's "lifeblood," and have risen to become its third or fourth largest operational expense (see Exhibit 5).

The Telecommunications Industry Association predicts that spending by U.S. corporations for total telecommunications will grow by an annual rate of 11.9 percent, reaching an all-time high of nearly $954 billions by 2004.[5] The cost for equipment and services will grow steadily as companies adopt new technologies. The actual cost of services will most likely keep going down as competitive pressures increase. However, the cost of *managing* communications will virtually double during the same period. According to another study by COMPASS America, an independent management consulting group, 30 percent of the total amount spent on telecommunications today goes to the in-house personnel that businesses need to manage communications equipment and services.[6] While those figures are sobering enough, the real cost of communications services is even greater if they are undermanaged. The burden on individual businesses is often hidden, always substantial, and extends into every area of the organization. At first, executives may notice seemingly unrelated issues of performance and capacity. Managers keep requesting more headcount. New facilities never seem to open on schedule. Business units are not in synch with each other.

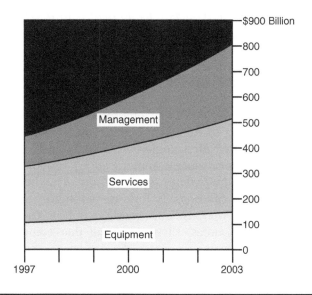

Exhibit 5. Total Cost of Communications

Early infrastructure problems are sometimes difficult to detect, but the costs of undermanaged communications are very real. For example:

- What is the cost of not opening a new location on time due to telecommunications delays?
- What is the consequence of having the wrong products or wrong configuration in place?
- What is the cost of lost business when data networks can't scale to accommodate customer demand?
- What is the cost when new employees lack the right communications tools when they report to work, particularly in a volatile labor market where the turnover may be as high as 20 to 40 percent, and where 455,000 technical jobs will go unfilled this year?[7]
- How many invoices are you receiving per month? Is anyone really managing these invoices? And what is the cost to your company when critical financial insights are missed because proper analysis is so prohibitively difficult?

The Communications Management Challenge

The truth is, communications is not managed as well as it could be, and for understandable reasons. First of all, the industry has changed more in the last five years than it did in its first hundred years. Second, high-growth businesses place severe demands on the communications infrastructure. Executives now have to struggle with a series of tough issues that were unknown a decade ago:

- *Communications technology is advancing at a dizzying pace.* With the continual introduction of new wireless, data, and broadband technologies, it becomes increasingly difficult for telecom or IT managers to stay current with the latest trends.

■ *Telecommunications managers have to deal with a staggering number of vendors in order to fulfill their needs worldwide.* What was once an easy decision between AT&T, MCI, Sprint, or WorldCom has ballooned to over 60 long-distance providers. Add to that the number of local providers, Internet providers, data network providers, wireless providers, and multiply by the number of market offers per provider for a mishmash of over 40,000 different offers to evaluate. Large enterprises may have to deal with as many as 100 vendors to fulfill all their needs at all their locations. Even for smaller businesses with a single corporate location, it's not uncommon to have a half dozen or more communications vendors.

■ *Service options are multiplying exponentially.* The best combination of carriers and service offerings is lurking out there, somewhere, buried among those 40,000 options. The question is how can companies, with limited staff and resources, sift through all that complexity to find the perfect solution?

■ *Frequent mergers and acquisitions are continually transforming a company's voice and data infrastructure.* Mergers and acquisitions have become a way of life for high-growth companies. They vastly increase the complexity of communications issues by adding new facilities virtually over night. IT managers suddenly become responsible for offices that are located in different states, different time zones, subject to different vendors, regulations, and billing procedures.

■ *IT departments are finding it difficult to build the communications infrastructure fast enough to support the growth goals of their company.* IT managers are facing a growing number of employees and locations. End-user demand increases as time-to-market pressures increase. As IT departments become inundated with basic provisioning, maintenance, and upgrades, they become the bottleneck within the company. Their focus is shifted away from core issues, such as ensuring maximum network uptime 24 × 7. Other departments and operating units are tempted to go around them, thereby increasing the fragmentation of the acquisition and management of the communications services. Industry trends such as these can quickly outpace the resources businesses have to manage them. In today's uncertain, highly competitive economy, additional headcount or budget is no longer the solution. Since communications has become mission-critical infrastructure, enabling virtually every business process, it needs to be managed as such. As companies attempt to sustain growth or just maintain parity with the competition, they are willing to redefine the rules of business and seek new, innovative solutions to communications problems. Companies that have relied on powerful managed solutions to gain control over human resources, accounting, sales, and manufacturing are realizing it's time to apply the same principle to communications services.

Part Two: The Solution: Managed Communications

A 1999 *Economist* report entitled "Vision 2010: New Strategies for Communications Enterprises,"[8] shows that communications providers are responding to the communication challenge in three different ways:

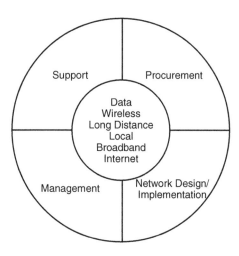

Exhibit 6.   Managed Communications Solution

- *Specialized carriers.* Some traditional providers are choosing to focus on a particular service, be it broadband or wireless. By keeping their focus narrow, they can devote all their resources to building the geographical reach and efficiencies of scale necessary to offer inexpensive service.
- *Integrated carriers.* Companies such as AT&T or XO are expanding their areas of operation to nearly all businesses in the communications industry, from traditional phone service to Internet access and cable. Companies turn to integrated carriers for the reduction in cost of ownership and convenience they afford. They also enable companies to implement new communications services more easily.
- *Managed communications.* The third model represents the greatest departure from the traditional telecommunications business. These providers act as integrators, assembling the best services into an integrated, customized solution that can be easily managed by a client's IT and finance departments. And, because they aren't burdened by owning the network, they are more nimble than specialized or integrated carriers. Providers of managed communications can choose the very best carrier for each service, and even switch partners if a better one emerges. They also offer infrastructure management and other value added processes, providing a full-service outsourced solution.

A managed communications solution can provide a single point of contact for four basic services (see Exhibit 6):

- *Procurement.* Whether you need to add one user locally, or are looking to integrate remote offices into your existing infrastructure, you can turn to one source.
- *Network design and implementation.* More than a communications carrier, a provider of managed voice and data solutions can work with you to design and implement a voice and data network that meets the growing needs of your business.

■ *Management.* Optimally, a full-service solution provides a set of financials tools for administering bills, validating expenses, allocating costs, reporting, and planning.

■ *Support.* Perhaps the main advantage of working with only one provider is that you get a single point of contact. No longer do you need to deal with conflicting vendors when network issues arise. Updating equipment and services also becomes faster and easier.

Part Three: The Benefits of Managed Communications

The trend toward strategic outsourcing is well documented in the business literature. Outsourcing is an accepted and fundamental strategy for successfully competing in today's economy. Outsourcing to providers of managed communications offers many benefits to a growing enterprise:

■ Access to world-class capabilities
■ Quick reengineering of business processes
■ Cost containment or reduction
■ Improved business focus
■ Resources redirected to more strategic activities
■ Control of difficult-to-manage functions
■ Increased flexibility and responsiveness
■ A hedge against the growing shortage of skilled, high-tech workers

Applied to communications services, these benefits translate into significant gains in efficiency and cost reductions for your company.

■ *Greater management control.* Experience has shown that managed communications is the most effective way to control communications services. By consolidating vendors, services, and billing, executives and managers finally get a complete and coherent overview of their usage and costs. Unnecessary services are no longer hidden amid the confusing array of multiple vendors who are often managed by different operating units within your company. Most importantly, management decisions can be based upon real-time data.

■ *Cost reduction and containment.* Because a full-service solution can help you manage your communications services more efficiently, it also reduces your costs significantly. Some companies, like Broadbase Software, see the total cost of communications services cut by 25 percent, with administrative overhead cut by as much as 75 percent.[9]

■ *Automation of administrative tasks.* Once a solution is in place, IT managers discover that they can finally *manage.* Freed from the repetitive administrative chores, they can focus on broader business issues, long-range planning, and ensure that the communications infrastructure supports the growth of the enterprise.

■ *Best-of-breed communications services.* Providers offering full-service, managed solutions maintain long-lasting relationships with a wide range of carriers. Vendors are put through a rigorous evaluation process and continually monitored to enforce standards they establish.

- *Flexibility to address change and growth.* When dealing directly with carriers, businesses often get locked into inflexible contracts. That doesn't happen when working with a provider of managed communications. IT managers have more flexibility to address regional or national requirements, and to quickly scale communications services as growth demands.
- *Better customer service.* Because they provide a single source of service, managed providers can offer better service and increased accountability. They have the capabilities to keep you better informed about every aspect of your mission-critical communications infrastructure.
- *Refocusing on core competencies.* Outsourcing of critical, but non-core functions enables companies to free resources to focus on core issues, thereby increasing individual productivity and operational efficiency.

Notes

1. *Fortune Magazine,* "Value-Driven Customer-Focused Outsourcing" at www.fortune.com/fortune/sections/outsourcing/outsourcing.htm.
2. Forrester Research, Forrester Report: "Buying into Telecom Online," July 2000.
3. *Fortune Magazine,* ibid.
4. AMI-Partners New York, "Medium Business Internet, E-Business & Telephony" July 2000.
5. "2001 MultiMedia Telecommunications Market Review and Forecast: Key Findings and Forecasts through 2004," a report of the Telecommunications Industry Association.
6. "Telecommunications Services Strategies: Think Globally or Act Locally," *Compass America,* 1998.
7. IDC, 2001.
8. "Vision 2010: New Strategies for Communications Enterprises," *The Economist*, 1999.
9. Schafer, Scott, "Romancing the Phone," *California CEO*, February 2001.

Disclaimer

The Economist and *The Economist* logo are trademarks of The Economist Newspaper Limited. All other names are trademarks or registered trademarks and are the property of their respective owners.

Appendix C

Summary of Communications Tax Regulations, United States Internal Revenue Service

Communications Tax

A 3 percent tax is imposed on amounts paid for all the following communications services:

- Local telephone service
- Toll telephone service
- Teletypewriter exchange service

Local Telephone Service

This means access to a local telephone system and the privilege of telephonic quality communication with most people who are part of the system. Local telephone service also includes any facility or services provided in connection with this service. The tax applies to lease payments for certain customer premises equipment (CPE) even though the lessor does not also provide access to a local telecommunications system.

Private Communication Service

Private communication service is not local telephone service. Private communication service includes accessory-type services provided in connection with a Centrex, PBX, or other similar system for dual-use accessory equipment. However, the charge for the service must be stated separately from the charge for the basic system, and the accessory must function, in whole or in part, in connection with intercommunication among the subscriber's stations.

Toll Telephone Service

This means a telephonic quality communication for which a toll is charged that varies with the distance and elapsed transmission time of each communication. The toll must be paid within the United States. It also includes a long-distance service that entitles the subscriber to make unlimited calls (sometimes limited as to the maximum number of hours) within a certain area for a flat charge. Microwave relay service used for the transmission of television programs and not for telephonic communication is not a toll telephone service.

Teletypewriter Exchange Service

This means access from a teletypewriter or other data station to a teletypewriter exchange system and the privilege of intercommunication by that station with most persons having teletypewriter or other data stations in the same exchange system.

Figuring the Tax

The tax is based on the sum of all charges for local or toll telephone service included in the bill. However, if the bill groups individual items for billing and tax purposes, the tax is based on the sum of the individual items within that group. The tax on the remaining items not included in any group is based on the charge for each item separately. Do not include in the tax base state or local sales or use taxes that are separately stated on the taxpayer's bill.

If the tax on toll telephone service is paid by inserting coins in *coin-operated telephones*, figure the tax to the nearest multiple of 5 cents. When the tax is midway between 5-cent multiples, the next higher multiple applies.

Prepaid Telephone Cards

A prepaid telephone card is any card or any other similar arrangement that allows its holder to get local or toll telephone service and pay for those services in advance. The tax is imposed when the card is transferred by a telecommunications carrier to any person who is not a telecommunications carrier. The face amount of the card is the amount paid for communications services. If the face amount is not a dollar amount, see Section 49.4251-4 of the Internal Revenue Service regulations.

Exemptions

Payments for certain services or payments from certain users are exempt from the communications tax.

Installation Charges

The tax does not apply to payments received for the installation of any instrument, wire, pole, switchboard, apparatus, or equipment. The tax does apply to payments for the repair or replacement of those items, incidental to ordinary maintenance.

Answering Services

The tax does not apply to amounts paid for a private line, an answering service, or a one-way paging or message service if they do not provide access to a local telephone system and the privilege of telephonic communication as part of the local telephone system.

Mobile Radio Telephone Service

The tax does not apply to payments for a two-way radio service that does not provide access to a local telephone system.

Coin-Operated Telephones

Payments made for services by inserting coins in coin-operated telephones available to the public are not subject to tax for local telephone service. They also are not subject to tax for toll telephone service if the charge is less than 25 cents. But the tax applies if the coin-operated telephone service is furnished for a guaranteed amount. Figure the tax on the amount paid under the guarantee, plus any fixed monthly or other periodic charge.

Telephone-Operated Security Systems

The tax does not apply to amounts paid for telephones used only to originate calls to a limited number of telephone stations for security entry into a building. In addition, the tax does not apply to any amounts paid for rented communication equipment used in the security system.

News Services and Radio Broadcasts of News and Sporting Events

The tax on toll telephone service and teletypewriter exchange service does not apply to news services and radio broadcasts of news and sporting events. The tax does not apply to charges for the following services.

- Services dealing exclusively with the collection or dissemination of news for or through the public press or radio or television broadcasting
- Services used exclusively in the collection or dissemination of news by a news ticker service furnishing a general news service similar to that of the public press

This exemption applies to payments received for messages from one member of the news media to another member (or to or from their bona fide correspondents). For the exemption to apply, the charge for these services must be billed in writing to the person paying for the service and that person must certify in writing that the services are used for an exempt purpose.

Services Not Exempted

The tax applies to amounts paid by members of the news media for local telephone service. Toll telephone service in connection with celebrities or special guests on talk shows is subject to the tax.

Common Carriers and Communications Companies

The tax on toll telephone service does not apply to WATS (wide area telephone service) used by common carriers, telephone and telegraph companies, or radio broadcasting stations or networks in their business. A common carrier is one holding itself out to the public as engaged in the business of transportation of persons or property for compensation and offering its services to the public generally.

Military Personnel Serving in a Combat Zone

The tax on toll telephone services does not apply to telephone calls originating in a combat zone that are made by members of the U.S. Armed Forces serving there if the person receiving payment for the call receives a properly executed certificate of exemption. The signed and dated exemption certificate must contain all the following information:

- The name of the member of the U.S. Armed Forces performing services in the combat zone who originated the call
- The toll charges, point of origin, and name of carrier
- A statement that the charges are exempt from tax under Section 4253(d) of the Internal Revenue Code
- The name and address of the telephone subscriber

This exemption also applies to members of the Armed Forces serving in a qualified hazardous duty area. A qualified hazardous duty area is either of the following areas:

- Bosnia and Herzegovina, Croatia, or Macedonia, effective November 21, 1995
- Federal Republic of Yugoslavia (Serbia/Montenegro), Albania, the Adriatic Sea, and the Ionian Sea north of the 39th parallel, effective March 24, 1999

A qualified hazardous duty area includes an area only while the special pay provision is in effect for that area.

International Organizations and the American Red Cross

The tax does not apply to communication services furnished to an international organization or to the American National Red Cross.

Nonprofit Hospitals

The tax does not apply to telephone services furnished to income tax-exempt nonprofit hospitals for their use. Also, the tax does not apply to amounts paid by these hospitals to provide local telephone service in the homes of its personnel who must be reached during their off-duty hours.

Nonprofit Educational Organizations

The tax does not apply to payments received for services and facilities furnished to a nonprofit educational organization for its use. A nonprofit educational organization is one that satisfies all the following requirements:

- It normally maintains a regular faculty and curriculum.
- It normally has a regularly enrolled body of pupils or students in attendance at the place where its educational activities are regularly carried on.
- It is exempt from income tax under section 501(a) of the Internal Revenue Code.

This includes a school operated by an organization that is exempt under Section 501(c)3 of the Internal Revenue Code if the school meets the above qualifications.

Federal, State, and Local Governments

The tax does not apply to communication services provided to the government of the United States, the government of any state or its political subdivisions, the District of Columbia, or the United Nations. Treat an *Indian tribal government* as a state for the exemption from the communications tax only if the services involve the exercise of an essential tribal government function.

Exemption Certificate

Any form of exemption certificate will be acceptable if it includes all the information required by the Internal Revenue Code and Regulations. File the certificate with the provider of the communication services.

The following users that are exempt from the communications tax do not have to file an annual exemption certificate *after* they have filed the initial certificate of exemption from the communications tax:

- The American National Red Cross and other international organizations
- Nonprofit hospitals
- Nonprofit educational organizations
- State and local governments

The federal government does not have to file any exemption certificate. All other organizations must furnish exemption certificates when required.

Credits or Refunds

If tax is collected and paid over for certain services or users exempt from the communications tax, the collector may claim a credit or refund if it has repaid the tax to the person from whom the tax was collected or obtained the consent of that person to the allowance of the credit or refund. Alternatively, the person who paid the tax can claim a refund.

Note

1. This document is available at http://www.irs.gov/plain/forms_pubs/pubs/p51004. htm. Some information in this document regarding air transportation taxes was omitted because it was not relevant to telecommunications cost management.

Appendix D

Web Sites of Interest

Broadband

DSL information — www.dsl.com/

Cable

Cable Labs (cable technology) — www.cablelabs.com/
Cable modem information — www.cable-modem.net/

Call Accounting

Call accounting information — www.telephonereporting.com/

Call Centers

Siebel Call Center Technologies — www.siebel.com/products/service/
 call_center/index.shtm
White Papers and Best Practices — www.call-center.net/

CLEC

AT&T — www.att.com/
MCI — www.mci.com/index.jsp

Cost Management

TSL (PRG) — www.itleverage.com/research/whitepapers/white_infra_
acct.html (White papers including VoIP chargeback)

General

International Telecommunications Union — www.itu.int/home/index.html
University of Michigan — http://china.si.umich.edu/telecom/
telecom-info.html

IXC

AT&T — www.att.com/
British Telecom — www.bt.com/index.jsp
MCI — www.mci.com/index.jsp

LEC

British Telecom — www.bt.com/index.jsp
Focal Communications — www.focal.com

Outsource

Profitline (telecom management) — www.profitline.com/
QuantumShift — www.quantumshift.com/
Telegistics — www.telegistics.com/

PBX

Avaya — www.avaya.com/
Ericsson — www.ericsson.nl/
Intel (telephony basics) — www.intel.com/network/csp/solutions/ipt/
IP Telephony On-line — www.internet-telephony.org/
Office Telephones — www.thephonesource.com/office.htm

Security/Privacy

Electronic Frontier Foundation — www.eff.org/
SecureLogix — www.securelogix.com

WAN

ATM Forum — www.atmforum.com/
British Telecom — www.bt.com/index.jsp
Frame Relay Forum — www.frforum.com/

Wireless

WAP Forum — www.wapforum.org/

US Telecom Association Members

@Comm Corporation — www.atcomm.com
3Com. orporation — www.3com.com
4–1–1 Systems, Inc. — www.411-systems.com
A&A Services — www.aaservices.com
Abacon Telecommunications — www.abacontel.com
Abiliti Solutions — www.abilitisolutions.com
Accelerated Networks — www.acceleratednetworks.com
AccuDocs LLC — www.accudocs.com
Actelis Networks — www.actelis.com
Acumen Solutions — www.acumensolutions.com
ADC — www.adc.com
ADTRAN, Inc. — www.adtran.com
AFNI, Inc. — www.afninet.com
AG Communications — www.agcs.com (ETC Member)
Alliance for Telecommunications Industry Solutions — www.atis.org
Alloptic, Inc. — www.alloptic.com
Altec Industries, Inc. — www.altec.com
AMDOCS, Inc. — www.amdocs.com
American Management Systems, Inc. — www.amsinc.com
Ameritraining, Inc. — www.ameritraining.com
Applied Innovation Inc. — www.aiinet.com
Aptis, Inc. — www.aptissoftware.com
Arguss Communications Group — www.arguss.com
Arista Information Systems, Inc. — www.aristainfo.com
ASAP Telecommunications Services, Inc. — www.asapintl.com
Astound Incorporated — www.astound.com
ATOGA Systems, Inc. — www.atoga.com
BB&T Capital Markets — www.scottstringfellow.com
Beacon Telecommunications Advisors, LLC — www.beaconbright.com
Bear, Stearns & Company, Inc. — www.bear.com
Bennet and Bennet, PLLC — www.bennetlaw.com
Berry Company (The) — www.lmberry.com
Berry, Dunn, McNeil & Parker — www.bdmp.com
BIA Capital Corp. — www.bia.com

Billing Concepts — www.billingconcepts.com

Blooston Mordkofsky Dickens Duffy & Prendergast —
　　www.bloostonlaw.com

Bond & Pecaro, Inc. — www.bondpecaro.com

Boston Communications Group — www.bcgi.net

BroadVision — www.broadvision.com

Brown Brothers Harriman & Co. — www.BBH.com

Bull HN Information Systems, Inc. — www.us.bull.com/telecom

Burton Training Group, Inc. — www.burtontraining.com (ETC Member)

CA-BOTICS Fiber Systems — www.ca-botics.com

Calix, Inc. — www.calix.com

CallVision — www.callvision.com

Carrier Access Corporation — www.carrieraccess.com

CashPoint Network Services, Inc. — www.cashpoint.net

Catena Networks — www.catena.com

CCG Consulting, Inc. — www.c-c-g.com

CCMI — www.ccmi.com

CDI Telecommunications, Inc. — www.cdicorp.com

CGI — www.cgiusa.com

CHR Solutions — www.chrsolutions.com

CIENA Corporation — www.ciena.com

Citicorp Diners Club — www.citibank.com/dinersus

CityNet Telecommunications, Inc. — www.citynettelecom.com
　　(ETC Member)

Cognitronics Corporation — www.cognitronics.com

COLO.COM — www.colo.com

CoManage Corporation — www.comanage.net

CommSoft — www.commsoft.net

Communications Data Group, Inc. — www.cdg-hargray.com

Comverse/Startel — www.startelcorp.com

congruency — www.congruency.com

Connexn Technologies, Inc. — www.cnnxn.com

Convergent Networks — www.convergentnet.com

Convergys — www.convergys.com

CON-X Corporation — www.con-x.com

Cooper, White & Cooper — www.cwclaw.com

Copper Mountain Networks — www.coppermountain.com

CopperCom — www.coppercom.com

CORE Telecom Systems, LLC — www.coretelecom.net

CoreNet Network Services, Inc. — www.corenetservices.com

Creative Support Solutions — www.solutionsbycss.com

CTI Group — www.ctigroup.com

Cummins Power Generation — www.CumminsPowerGeneration.com

CyberPath, Inc. — www.cyberpathinc.com

Cygent, Inc. — www.cygent.com

Daniels & Associates — www.bdaniels.com

Dantel, Inc. — www.dantel.com

Data Center, Inc. — www.datacenterinc.com
dB TELCO, Inc. — www.dbtelco.com
DETECON, Inc. — www.deteconusa.com
docHarbor — www.docharbor.net
DukeNet Communications — www.duke-energy.com
Eastern Research, Inc. — www.erinc.com
ECI Telecom, Inc. — www.ecitele.com
EHPT USA Inc. — www.ehptus.com
Eicon Networks — www.eicon.com
Elastic Networks — www.elastic.com
Electronic Tele-Communications www. — etcia.com
Emergia — www.e-mergia.com
eMIS — eMerging Information Systems — www.emis-intl.com
Energy Technologies, Inc. — www.powersource.net
Engel Consulting Group — www.engelgroup.com
EUR Systems, Inc. — www.eursystems.com
Evolving Systems, Inc. — www.evolving.com
Exigen — www.exigengroup.com
Expertech Network Installation, Inc. — www.expertech.net
Falkenberg Capital Corporation — www.falkenbergcapital.com
FAT PIPE Magazine — www.fatpipeonline.com
FileTek, Inc. — www.filetek.com
First Marketing — www.first-marketing.com
Fletcher, Heald & Hildreth — www.fhh-telcomlaw.com
FPTA Technologies, Inc. — www.fpta.com
Fred Williamson & Associates, Inc. — www.fwainc.com
Fujitsu Network Services — www.fns.fujitsu.com (ETC Member)
Galactech Corporation — www.galactech-corp.com
GE Capital Vendor Financial Services — www.geleaseconnect.com
General Bandwidth — www.genband.com
General Cable Corporation — www.generalcable.com
Gluon Networks — www.gluonnetworks.com
GNP Computers, Inc. — www.gnp.com
GoDigital Networks — www.godigital.com
Gordon Parker — gordon@gordonparker.com
Gryphon Networks — www.gryphonnetworks.com
GVNW Consulting, Inc. — www.gvnw.com
Harris Corporation — www.harris.com
Harte-Hanks — www.harte-hanks.com
Henkels & McCoy, Inc. — www.henkelsandmccoy.com
Hitachi Telecom (USA), Inc. — www.hitel.com
HNC Software, Inc. — www.hnc.com
HOMISCO/VOICENET — www.homisco.com
HyperEdge Corp. — www.hyperedge.com
ICORE, Inc. — www.icoreinc.com
IDN, L.L.C. — www.idnllc.net
ILD Telecommunications Inc. — www.ildtelecommunications.com

Illuminet — www.illuminet.com
iMagic TV, Inc. — www.imagictv.com
Independent NECA Services, Inc. — www.ins.org
Industrial Logic Corporation (ILC) — www.indlogic.com
Info Avenue Internet Services, LLC — www.infoave.net
Information Intellect, Inc. — www.infointellect.com
Innovative Systems, LLC — www.innovsys.com
Intec Telecom Systems — www.intec-telecom-systems.com
Integral Access — www.integralaccess.com
Intercept Compliance Solutions, LLC — www.interceptcompliance.com
Inventronics Limited — www.inventronics.com
Iowa Network Services Inc. — www.netins.net
Issue Dynamics, Inc — www.idi.net
Jackson Thornton & Co. (CPAs) — www.jacksonthornton.com
Jeffries & Company, Inc. — www.jefco.com
John Staurulakis, Inc. — www.jsitel.com
Johnson, Stone & Pagano — www.jspcpa.com
Joslyn Electronic Systems Company — www.jesc.com
JSI Capital Advisors, LLC — www.jsicapital.com
Kaparel Corporation — www.kaparel.com
Kiesling Associates LLP — www.kiesling.com
KPMG Corporate Finance LLC — www.kpmg.com
KPMG Peat Marwick LLP — www.kpmg.com
LAN Hopper Systems, Inc. — www.lanhopper.com
Lasertech — www.lasertechinc.com
Latham and Watkins — www.lw.com
Lawson Software — www.lawson.com
League Technologies — www.leaguetech.com
Legg Mason Wood Walker, Inc. — www.leggmason.com
Lexent Inc. — www.lexent.net
Lightbridge, Inc. — www.lightbridge.com
Linc.net, Inc. — www.linc.net
Listing Services Solutions Inc. — www.lssi.net
Loop Expert Technologies — www.loopexpert.com
Lukas, McGowan, Nace & Gutierrez — www.fcclaw.com
LuxN, Inc. — www.luxn.com
LuxVue, Inc. — www.luxvue.com
MAPCOM Systems, Inc. — www.mapcom.com
Mariner Networks — www.marinernetworks.com
MarotzTel, Inc. — www.marotztel.com
Marsh USA, Inc. — www.Marshweb.com
Martin Group — www.martin-group.com
Mazer Telecom Advisors, LLC — www.mazertel.com
McCall-Thomas Engineering Company Inc. — www.mcteng.com
McCartney & Company, P.C. — www.mcco-cpa.com
McLean & Brown — www.mcleanbrown.com

MDSI Mobile Data Solutions, Inc. — www.mdsi-advantex.com (ETC Member)

Merlot Communications — www.merlotcom.com

MetaSolv Software Inc. — www.metasolv.com

MetraTech Corp. — www.metratech.com

Metrobility Optical Systems — www.metrobility.com

Michael Baker Jr., Inc. — www.mbakercorp.com

Microsoft Corporation — www.microsoft.com

Mid America Computer Corporation (MACC) — www.maccnet.com

Mid-South Consulting Engineers — www.msceng.com

Mid-State Consultants, Inc. — www.mscon.com

Midwest Communication Products, Inc. — www.groupmidwest.com

Minerva Networks — www.minervanetworks.com

Mirapoint, Inc. — www.mirapoint.com

Moss Adams LLP — www.mossadams.com

MPhase Technologies, Inc. — www.mPhaseTech.com

Myrio Corporation — www.myrio.com

National Emergency Number Association — www.nena.org

National Exchange Carrier Association (NECA) — www.neca.org

National Information Solutions Cooperative (NISC) — www.nisc.cc

National Rural Telecommunications Co. — www.nrtc.org

Nations Media Partners, Inc. — www.nationsmedia.com

Navigant Consulting, Inc. — www.pcit.com

NCR — www.ncr.com

Net to Net Technologies — www.nettonettech.com

Net-Hopper Systems, Inc. — www.net-hopper.com

Netigy Corporation — www.netigy.com

Network Services Group LLC — www.networkservicesgrp.com

NeuStar, Inc. — www.neustar.com

Nevik Networks — www.neviknetworks.com

NightFire Software, Inc. — www.nightfire.com

NTSG — www.ntsg.net

OAN Services, Inc. — www.oanservices.com

Occam Networks, Inc. — www.occamnetworks.com

Olsen Thielen & Company, Ltd. — www.olsen-thielen.com

Omniphone, Inc. — www.omniphoneinc.com

OnePath Networks Inc. — www.onepathnet.com (ETC Member)

Optical Solutions, Inc. — www.opticalsolutions.com

Orckit Communications — www.orckit.com

Oresis Communications — www.oresis.com

Orillion USA, Inc. — www.orillion.com

Orius Corporation — www.oriuscorp.com

Pace Micro Technology Americas — www.pacemicro.com

Paceon — www.paceon.com

Panasonic Personal Computer Company — www.panasonic.com/toughbook

PANDATEL AG — www.pandatel.com

Paradyne Corporation — www.paradyne.com
Parks Associates — www.parksassociates.com
Peco II, Inc. — www.peco2.com
Pelago Networks — www.pelagonet.com
PeopleSoft, Inc. — www.peoplesoft.com
Pirelli Cable Corporation — www.pirelli.com (ETC Member)
Pliant Systems — www.pliantsystems.com
Polaris Networks — www.polarisnetworks.com
Power & Telephone Supply Company — www.ptsupply.com
PRIMA, Inc. — www.prima.ca
Prince Market Research, Inc. — www.PMResearch.com
Proudfoot Consulting — www.proudfootconsulting.com
Pulsecom — www.pulse.com
Quintrex Data Systems Corporation — www.quintrex.com
Raviant Networks Inc. — www.raviant.com
Raze Technologies, Inc. — www.razetechnologies.com
Redback Networks — www.redback.com
Redcom Laboratories, Inc. — www.redcom.com
Resource Advantage, Inc. (RAI) — www.raiconsulting.com
Rivercode LLC — www.rivercode.com
Robbins-Gioia, Inc. — www.robbinsgioia.com
RVW, Inc. — www.rvwinc.com
Santera Systems — www.santera.com
Science Applications International Corporation (SAIC) —
    www.saic.com (ETC Member)
Secure Communications, Inc. — www.securecominc.com
Selectek — www.selectek.com
SGS International Certification Services, Inc. — www.sgsicsus.com
Signal Equity Partners — www.signal-equity.com
SM&P Utilities Resources, Inc. — www.sm-p.com
SOMA Networks, Inc. — www.somanetworks.com
Spark Interactive — www.sparkinteract.com
Sprint North Supply — www.sprintnorthsupply.com (ETC Member)
SPSS Inc. — www.spss.com (ETC Member)
Sterling Commerce — www.stercomm.com
Subex Systems Ltd. — www.subexgroup.com
Sun Microsystems — www.sun.com
SunTec Business Solutions (Pvt. Ltd.) — www.suntecgroup.com
Superior Telecommunications Inc. — www.superioressex.com
Suttle — www.suttleonline.com
SyTech Corporation — www.sytechcorp.com
T.Soft Inc. — www.tsoftus.com
TAMCO — www.tamcocorp.com
Taqua Systems — www.taqua.com (ETC Member)
Tarantella — www.tarantella.com
TARGUS Information Corporation — www.targusinfo.com
TCA Inc. — Telcom Consulting Associates — www.tcatel.com

Tekelec, Inc. — www.tekelec.com
Telamon Corporation — www.telamon-corp.com
Tele Resources, Inc. — www.teleresources.net
TELEC Consulting Resources, Inc. — www.telec-consulting.com
Telecom & Energy Cables Corp. — www.tecables.com
TeleCon L.T.D./COBRA Solutions, Inc. — www.teleconltd.com
Telect, Inc. — www.telect.com
Telerx — www.telerx.com
Telesphere — www.tele-sphere.com
Telica, Inc. — www.telica.com
Telution, Inc. — www.telution.com
Terawave Communications, Inc. — www.terawave.com
Terayon Communication Systems — www.terayon.com
The Billing College — www.billingcollege.com
The Mangement Network Group (TMNG) —
    www.tmng.com (ETC Member)
Tiara Networks — www.tiaranetworks.com
Top Layer Networks, Inc. — www.TopLayer.com (ETC Member)
Turin Networks, Inc. — www.turinnetworks.com
Turnstone Systems, Inc. — www.turnstone.com
TVC Communications — www.tvcinc.com
U.S. Connect — www.thereal411.com
UDP, Inc. — www.udpinc.com
Ultrapro International, Inc. — www.ultrapro.com
Underwood Sales Corporation — www.underwoodsales.com
Universal Access, Inc. — www.universalaccess.net
UniView Technologies — www.uniview.com
uReach — www.ureach.com
US Communications Technology, Inc. — www.usct.net
UtiliTec — www.utilitec.net
UTILX Corporation — www.utilx.com
Vertex, Inc. — www.vertexinc.com
ViaGate Technologies — www.viagate.com
VideoTele.com — www.videotele.com
Vitalwork Inc. — www.vitalwork.com
Voice Mobility Inc. — www.voicemobility.com
Warinner, Gesinger & Associates, LLC — www.wgacpa.com
WaveRider Communications, Inc. — www.waverider.com
WebTone Technologies — www.webtonetech.com
Westell Technologies, Inc. — www.westell.com
WhisperWire — www.whisperwire.com
White Rock Networks, Inc. — www.whiterocknetworks.com
Wiley, Rein & Fielding — www.wrf.com
Woltcom, Inc. — www.woltcom.com
World Wide Packets — www.worldwidepackets.com
ZCorum, Inc. — www.zcorum.com
Zhone Technologies — www.zhone.com

Appendix E

Sample RFP Template

The template below is a starting point only. It should be modified to fit the project and culture of the organization. Also, any major RFP should be reviewed by legal counsel to ensure that the organization's interests are appropriately guarded.

Cover Page

The cover page should include the company name, the type of services being bid on, and some type of document control number (e.g., year, month, and initials of point person)

Company Name
Local & Long-Distance Telephone Services
REQUEST FOR PROPOSAL
CONTROL NO.
RFP 2001–07-NN

Table of Contents

The document structure should fit the style of the organization. For example, all specifications and/or requirements may exist within one section or may be broken out into separate sections (see Exhibit 1).

Section I: RFP Response Information

Provide a description of the company, with particular attention to drivers such as number of employees, locations, etc. Also, state the major objective and scope of the request. The following 12 subsections should be included or excluded as appropriate.

Exhibit 1. Sample Outline

Section I	RFP Response Information
Section II	Contract Terms and Conditions
Section III	Specifications/Requirements
Section IV	Proposal Information, Cost, and Pricing

1. Proposal/No Bid Submission

All proposals in response to this RFP shall be received no later than 5:00 p.m., March 15, 2002. Please respond with your intent to submit a proposal to this RFP, using the attached "Declaration of Intent to Quote" as soon as possible to the e-mail address below, but no later than the closing date of March 09, 2002. Responses may be submitted electronically by the above date. A hard copy must arrive by the following day, March 16, 2002.

To: e-mail address of person within the company

2. Request for Additional Information

Telephone inquiries will not be accommodated unless all questions are directed to *<Company Procurement Department Name>*. Any questions regarding this RFP should be directed to *<Contact Name/RFP Issuer>*, at *<(555) 111–2222>*, or e-mail to *<Address>*. Verbal information or explanation given shall not be binding on *<COMPANY>*. E-mail is preferable.

A Bidder's Teleconference will be held on March 12, 2002, at 3:00 p.m. EST. The dial-in number is as follows: Dial In: (555) 111–2222 Participant Code: 111111.

3. Acceptance Period

Proposals in response to this RFP shall indicate that they are valid for a period of not less than three (3) months from the closing date.

4. Contract Award

Company shall make the contract award in connection with this RFP based on our evaluation of your technical and price proposal. Company shall not construe this RFP as any commitment. Company reserves the right to accept any proposal or reject all proposals. Company also reserves the right to negotiate any proposal after receipt. Final selection of the supplier does not constitute acceptance of the submitted proposal.

5. Notification

Notification of the award will be made upon completion of the selection process. Suppliers not being awarded this contract will be notified in writing.

6. Contract

Any award as a result of this RFP shall be contingent upon the execution of an appropriate contract.

Section II of this RFP contains our proposed Terms and Conditions. These Terms and Conditions shall form the basis of a Contract covering the subject matter of this RFP. The Acceptance Form found in Section II must accompany proposals submitted hereunder. If there is contention(s) with the Terms and Conditions, a brief explanation and alternative language, if any, should be included on the Acceptance Form. Any exceptions to the Terms and Conditions will be taken into consideration when evaluating proposals submitted. Company reserves the right to reject any or all of your proposed modifications.

7. Use of Information
This RFP and any other information furnished under this RFP shall be used solely for the purposes of responding to this RFP. All such documents and information shall remain the property of Company, shall be kept confidential and shall be returned to Company upon request. Reproduction of any part of this RFP is authorized only to the extent necessary for preparation of your response; all such copies are to be destroyed when no longer required in connection with this RFP.

Any information provided to Company as a result of this RFP shall not be considered by Supplier to be confidential or proprietary.

8. Pricing
Pricing shall be in the format contained in Section IV: Proposal Information, Cost, and Pricing of this RFP. Alternative approaches for the pricing of the requested services may be provided; however, such alternate approaches shall be described separately and must be in addition to the format in Section IV. All proposals must include anticipated expenses, such as travel, and proposed budget. Do not include cost or price figures anywhere except in the cost and pricing section.

9. Contacts
During our proposal evaluation period, questions regarding your proposal may arise. Your cover letter should include both the name(s) and phone number(s) of the individual(s) who should be contacted by <COMPANY> as such questions arise or should a presentation be requested.

10. Amendments
In the event it becomes necessary to revise, change, modify, or cancel this RFP or to provide additional information, an amendment will be issued to all recipients of this RFP.

11. Alternative Responses
It is Company intent to solicit quotations that afford the most cost-efficient, technically responsible proposals for these services. However, Company recognizes there may be arrangements different from that requested hereunder that would offer additional benefits while satisfying the applicable specifications. Accordingly, you may submit alternative proposals for consideration, which offer such additional benefits in addition to the requested pricing

information. These alternatives shall be evaluated in conjunction with the data provided for each proposal.

12. Unnecessary Elaborate Proposals
You are requested not to make your proposal elaborate with respect to presentation. A simple, straightforward, economically reproduced proposal is strongly recommended. Company's proposal evaluation procedure places a higher premium on thoroughness of presentation (i.e., responsiveness) rather than on quantity of material included.

Section II: Contract Terms and Conditions

In many cases, the RFP will ask the bidder to provide a proposed purchase and sale agreement. In the telecom world, the latter is a common practice. Laws change; all clauses should be reviewed by legal counsel and the procurement group. While most organizations use standard, in-house terms and conditions for small to mid-sized projects, more complex language may be required for large projects. An example of contract terms and conditions is provided at the end of this appendix.

Sample General Requirements

Vendor shall provide, as an attachment to this quote, a proposed Purchase and Sale Agreement. This shall provide the framework of the contract with the selected vendor. This contract shall address the following issues at a minimum:

- Reference to the vendor response to this RFP
- Pricing sheets from the RFP package
- Firm installation completion date
- Scope of work
- Customer responsibility
- Training responsibilities
- Remedies
- Total purchase price
- Payment terms
- System acceptance language (based on four-week acceptance period)
- Warranty coverage
- Warranty period
- Maintenance and repair response times
- Confidentiality
- Limitation of liability
- Default
- Termination

Exhibit 2. Network Design Requirements Tailored to Each Customer

In an actual RFP, this attachment would show network requirements specific to each organization. It should be as detailed as possible and include bandwidth, reliability, uptime, power, configuration, and other architectural elements as needed.

Section III: Specifications/Requirements

Most requests for proposal are company unique with respect to specifications and requirements. The following provides several overall areas for inclusion, followed by sample requirement worksheets for the vendor to follow.

Vendor Information Required

Submit a list of all account team members (name, position, and telephone number) who will be responsible for the system under proposal.

Contact	Telephone Number
Sales Representative	XXX-XXXX
Service Representative	XXX-XXXX
Project Manager	XXX-XXXX
Repair Technician	XXX-XXXX
Remote Repair Tech.	XXX-XXXX

- Provide proof of state and federal certification to provide services in the applicable region.
- Provide documentation that illustrates geographic coverage of facilities-based network services.
- Provide a list of customer references that includes customer name, contact name, and telephone number. References should have the similar services as those being proposed in this RFP response.

Network Design

See Exhibit 2 for the network design requirements.

Invoice Requirements

The following invoice capabilities are required:

- Summary billing
- Duplicate invoices
- Electronic CD ROM format

- Web-based invoice access
- Local line cost breakdown
- Summary of calls by service type
- Longest call summary reports
- Most expensive call reports
- International call summary
- Area code reports
- Toll-free area code reports
- Toll-free most frequently received numbers

Sample General Customer Service Requirements

- What is the repair telephone number?
- Where are the repair centers located and what are the hours of operation?
- Describe the repair process.
- Describe the escalation process and time frames.
- What are the response times for repair calls affecting local and long-distance services? Indicate any variations based on time of day, day of week, or holiday schedules.
- What is the telephone number for billing and general services, hours, and where are they located?

Sample Service Level Agreement Requirements

A service level agreement (SLA) must be included with the response for the services being proposed. The service level objective for each category should include the following:

- Definition of network outage
- Definition of trouble ticket
- Availability
- Mean time to repair (MTTR)
- Distinction between local and long-distance services
- Distinction between dedicated and switched services
- Documentation requirements

In addition, the vendor should:

- Provide calculations used to determine SLA
- Provide credit structure for SLA
- Indicate whether the following reports are available:
  - Trouble ticket summary report with total tickets, tickets closed within SLA, and average/mean time to resolution by priority
  - Trouble ticket detail reports with the following information:
    - Activity by location (to include summary report information)
    - Analyst performance (to include summary report information)

- Project Summary
- Date opened
- Project or milestone description
- Responsible party
- Priority
- Status
- Comments
- Due Date

General Pricing Requirements

- Provide pricing for each of the listed services on the attached pricing sheet. (If there is a price difference between locations, please identify.)
- Provide the contract term for the listed services. (No more than a two-year contract is desired by <Company>.)
- Provide the date and number of the federal and state tariffs that dictate pricing.

The following three items are strongly preferred:

- Pricing caps for all services for the duration of the contract term
- Agreement to allow pricing structure reviews at any point prior to the end of the contract term this will help ensure that services are being charged at market rates
- Non-recurring fees for all services identified on the pricing sheet

Vendor Agreement

As stated in this RFP, all answers and information provided as part of the vendor's bid response will become part of the Purchase and Sales Agreement. Indicate compliance by signing below:

Authorization of RFP Response

Company: _____

By: _____

Typed Name: _____

Title: _____

Date: _____

Sample Requirement Worksheets

Exhibit 3.   Location Pricing Sheet

Office/Site X
Complete the following information for each location:
Services:
Quantity & Service Available? (Y/N)
Install time frame:
One-time costs:
Recurring monthly costs:
Trunking configurations:
Loop start analog lines
Ground start trunks
Centrex — analog lines
Analog DID trunks
T1 — DID/DOD
ISDN PRI — DID/DOD4
ISDN BRI
Off-premise extensions
Foreign exchange circuits
Tie lines
Point to point
Toll-free services:
Toll free — switched services
Toll free — dedicated services
Area code routing
Enhanced toll-free routing
Service options:
DID number ranges
Teleconferencing
Central office voicemail
Call forward busy/no answer
Hunt group — series, circular
Caller ID/caller ID blocking
Account codes — validated, unvalidated

Exhibit 4. Usage Pricing Sheet — All Locations

Usage type:
Initial billing increment
Ongoing billing increment
Cost per call
Rate per minute
Other cost outbound calling:
Switched origination
Local calls
IntraLATA (AB locations)
IntraLATA (CD locations)
InterLATA, IntraState (EF)
InterLATA, IntraState (HI)
Interstate
Directory assistance
Dedicated origination:
Local calls
IntraLATA (AB locations)
IntraLATA (CD locations)
InterLATA, Intrastate (AB)
InterLATA, Intrastate (CD)
Interstate
Directory Assistance inbound calling:
Switched termination:
Local Calls
IntraLATA (AB locations)
IntraLATA (CD locations)
InterLATA, Intrastate (AB)
InterLATA, Intrastate (CD)
Interstate
Dedicated Termination:
Local Calls
IntraLATA (AB locations)
IntraLATA (CD locations)
InterLATA, Intrastate (AB)
InterLATA, Intrastate (CD)
Interstate

An inventory of sites and circuits is important for a pricing response. Provide the physical addresses of the sites as well as T1 and T3 circuit data. That is, include circuit information for location A to location B, type of circuit (e.g., point to point, PRI), the purpose of the circuit, and any comments such as what equipment the circuit terminates on or N-digit Inter-Office Dialing. Other detailed information such as a voice network diagram would be helpful.

Exhibit 5.   Proposal Table of Contents

Section 1	Executive/Proposal Summary
Section 2	Technical Capability and Quality Program
Section 3	Cost and Price Proposal
Section 4	Contract Proposal
Section 5	Additional Proposals

Section IV: Proposal Information, Cost, and Pricing

Introduction

This guide is to be used in the preparation and organization of your proposal. Your careful adherence to these instructions will enable us to perform a full, objective, and uniform evaluation.

Your proposal shall consist of the separate sections as shown in Exhibit 5.

Do not include cost or price figures anywhere except in the Cost and Price Proposal.

All proposals meeting the stated requirements and specifications, except for minor exceptions and deviations, shall be considered. Failure to meet requirements may disqualify a proposal from the selection process. However, proposals having minor exceptions and deviations shall be considered only if the following conditions are satisfied: (1) all exceptions and deviations from the specifications are explicitly stated in the Proposal Summary and fully detailed in the Technical Capability and Quality Program Section; and (2) all exceptions and deviations are appropriately justified on the basis of performance or relative price.

The criteria to be used for the proposal evaluation includes:

- Cost
- Technical merit
- Responsiveness to contract provisions
- Quality
- Service
- Innovative solutions
- Ease of doing business
- Documentation and reports
- Others as you deem appropriate

No weighing or relative importance of criteria is implied by the order of this list.

You shall furnish all information as requested and provide sufficient data to enable us to evaluate the proposal. Any deviations or exceptions to the RFP should be noted. Any respondent who does not completely reply to the proposal as requested may be eliminated at the discretion of Company.

The same article, section, or paragraph number and title used in the RFP shall be used for your comments.

In the cases where your reply is "will not be complied with" or "not agreed to," you shall indicate your reasons for such disagreement and provide an alternative with which you will comply or agree. In such instances, the wording in question should be cross-hatched out (i.e., xxxxxx) and the suggested language *added and underlined.*

Proposal Content

The required content of each section of your proposal follows.

Section 1: Executive/Proposal Summary

This section should provide a concise summary of the proposal and a summary of the key features of your proposal response. All deviations and exceptions from the RFP should be stated in short form and a brief justification given. This section should not exceed ten pages, and should *not* contain any cost or price information.

Section 2: Technical Capability and Quality Program

This section and the entire proposal shall not include resumes. If included, resumes will be removed without review.

This section shall include a detailed description of the service(s). In general, this section should address all requirements of Section 3 of this RFP.

Your response to this section shall detail the following:

- Account management skills, functions, and services provided by your firm
- Equipment, facilities, and staff to be used by your firm in the performance of services
- The Total Quality Management (TQM) or other quality program employed by your firm
- A description of the service capabilities of your firm and any detail on why it will be easy for Company to do business with your firm
- The capabilities of your firm to provide complete documentation and reports associated with your provision of service(s).

Section 3: Cost and Price Proposal

This section shall include a description of your proposed prices, including hourly rate schedules for specific levels of responsibility, proposed weekly hours and schedules, and anticipated expenses, if applicable. All pricing information shall be limited solely to this section of your proposal. This section should address all requirements set forth in Section 3, as well as any other items pertinent to your proposal. These requirements have been developed to allow us to uniformly evaluate prices submitted for the work.

Section 4: Contract Proposal

This section shall include your response to our proposed Terms and Conditions included in Section III, Contract Terms and Conditions, and shall form the basis for the preparation of a contract(s) covering the subject matter of this RFP.

You shall respond in your proposal either that all Terms and Conditions are acceptable or that some are acceptable and some are not. Underline or highlight those words, phrases, sentences, paragraphs, etc. that are not satisfactory and note any exceptions by referencing the appropriate article number. Add a brief explanation and alternative language, if any, and submit same on a separate typewritten sheet. Exceptions will be taken into consideration when evaluating your proposal.

Section 5: Additional Proposals

Sections 1 through 4 should address the requirements stated in this RFP. This section should address any *additional* proposals. Detailed discussion of the additional proposal should not be intermixed with your response to the specific requirements of this Request for Proposal. This means that any additional proposals you submit must be under separate cover.

Additional Questions

Bidder's teleconference: A project introduction and overview, as well as a question and answer period will be held on July 1, 2004, to accommodate questions. The date, time, and teleconference numbers will be announced.

Declaration of Intent to Quote

Supplier:
Address:
Supplier's Representative:
RFP/RFQ No.:
Company Representative:
Description of Service or Product:
Please check one of the following:
Our firm intends to submit a response to the above referenced RFP/RFQ.
Our firm DOES NOT intend to submit a response to the above referenced RFP/RFQ.
If you do not intend to respond, please answer the following:
Did the RFP/RFQ allow adequate response time?   Yes   No
Did the Statement of Work provide adequate detail in order to prepare a response?
  Yes   No
Is the product or service within your firm's area of business/expertise?   Yes   No
Please provide any additional information you may wish to share regarding your
  decision not to respond.

Supplier's Representative:

(Signature)
Date:
Fax to:

Appendix F

Sample Contract Terms and Conditions for a Complex Telecommunications Development Program

Note to reader: Any actual terms and conditions used by your organization should be reviewed by legal counsel.

Assignment and Subcontracting

Supplier shall not assign any right or interest under this Agreement (excepting monies due or to become due) or delegate or subcontract any Work or other obligation to be performed or owed under this Agreement without the prior written consent of Company. Any attempted assignment, delegation, or subcontracting in contravention of the above provisions shall be void and ineffective. Any assignment of monies shall be void and ineffective to the extent that (1) Supplier shall not have given Company at least thirty (30) days prior written notice of such assignment or (2) such assignment attempts to impose upon Company obligations to the assignee additional to the payment of such monies, or to preclude Company from dealing solely and directly with Supplier in all matters pertaining to this Agreement, including the negotiation of amendments or settlements of charges due. All Work performed by Supplier's subcontractor(s) at any tier shall be deemed Work performed by Supplier.

Changes

Company may at any time during the progress of the Work require additions to or alterations of or deductions or deviations (all hereinafter referred to as a "Change") from the Work called for by the specifications, drawings, and samples. No Change shall be considered as an addition or alteration to or deduction or deviation from the Work called for by the specifications, drawings, and samples nor shall Supplier be entitled to any compensation for work done pursuant to or in contemplation of a Change, unless made pursuant to a written Change Order issued by Company. Within ten (10) days after a request for a Change, Supplier shall submit a proposal to Company which includes any increases or decreases in Supplier's costs or changes in the delivery or Work schedule necessitated by the Change. Company shall, within ten (10) days of receipt of the proposal, either (1) accept the proposal, in which event Company shall issue a written Change Order directing Supplier to perform the Change or (2) advise Supplier not to perform the Change in which event Supplier shall proceed with the original Work.

Choice of Law

The construction, interpretation, and performance of this Agreement and all transactions under it shall be governed by the laws of the State of New Jersey, excluding its choice of laws rules and excluding the Convention for the International Sale of Goods. The parties agree that the provisions of the New Jersey Uniform Commercial Code apply to this Agreement and all transactions under it, including agreements and transactions relating to the furnishing of services, the lease or rental of equipment or material, and the license of software. Supplier agrees to submit to the jurisdiction of any court wherein an action is commenced against Company based on a claim for which Supplier has agreed to indemnify Company under this Agreement.

Compliance with Laws

Supplier and all persons furnished by Supplier shall comply at their own expense with all applicable federal, state, local, and foreign laws, ordinances, regulations, and codes, including those relating to the use of chlorofluorocarbons, and including the identification and procurement of required permits, certificates, licenses, insurance, approvals, and inspections in performance under this Agreement. Supplier agrees to indemnify, defend (at Company's request), and save harmless Company, its affiliates, its and their customers and each of their officers, directors, and employees from and against any losses, damages, claims, demands, suits, liabilities, fines, penalties, and expenses (including reasonable attorney's fees) that arise out of or result from any failure to do so.

Entire Agreement

This Agreement shall incorporate the typed or written provisions on Company's orders issued pursuant to this Agreement and shall constitute the entire agreement between the parties with respect to the subject matter of this Agreement and the order(s) and shall not be modified or rescinded, except by a writing signed by Supplier and Company. All references in these terms and conditions to this Agreement or to Work, services, material, equipment, products, software, or information furnished under, in performance of, pursuant to, or in contemplation of, this Agreement shall also apply to any orders issued pursuant to this Agreement. Printed provisions on the reverse side of Company's orders (except as specified otherwise in this Agreement) and all provisions on Supplier's forms shall be deemed deleted. Additional or different terms inserted in this Agreement by Supplier, or deletions thereto, whether by alterations, addenda, or otherwise, shall be of no force and effect, unless expressly consented to by Company in writing. Estimates or forecasts furnished by Company shall not constitute commitments. The provisions of this Agreement supersede all contemporaneous oral agreements and all prior oral and written quotations, communications, agreements, and understandings of the parties with respect to the subject matter of this Agreement. The term "Work" as used in this Agreement may also be referred to as "services."

Force Majeure

Neither party shall be held responsible for any delay or failure in performance of any part of this Agreement to the extent such delay or failure is caused by fire, flood, explosion, war, strike, embargo, government requirement, civil or military authority, act of God, or other similar causes beyond its control and without the fault or negligence of the delayed or nonperforming party or its subcontractors ("force majeure conditions"). Notwithstanding the foregoing, Supplier's liability for loss or damage to Company's material in Supplier's possession or control shall not be modified by this clause. If any force majeure condition occurs, the party delayed or unable to perform shall give immediate notice to the other party, stating the nature of the force majeure condition and any action being taken to avoid or minimize its effect. The party affected by the other's delay or inability to perform may elect to: (1) suspend this Agreement or an order for the duration of the force majeure condition and (a) at its option buy, sell, obtain, or furnish elsewhere material or services to be bought, sold, obtained, or furnished under this Agreement or an order (unless such sale or furnishing is prohibited under this Agreement) and deduct from any commitment the quantity bought, sold, obtained, or furnished or for which commitments have been made elsewhere, and (b) once the force majeure condition ceases, resume performance under this Agreement or an order with an option in the affected party to extend the period of this Agreement or order up to the length of time the force majeure condition endured; and/or (2) when the delay or nonperformance continues for a period of at least fifteen (15) days, terminate, at no charge, this Agreement or an order or the part of it relating to material not already shipped, or services not already performed. Unless written notice is given within forty-five (45) days after the affected party is notified of the force majeure condition, (1) shall be deemed selected.

Government Contract Provisions

The following provisions regarding equal opportunity, and all applicable laws, rules, regulations, and executive orders specifically related thereto, including applicable provisions and clauses from the Federal Acquisition Regulation and all supplements thereto are incorporated in this Agreement as they apply to work performed under specific U.S. Government contracts: 41 CFR 60–1.4, Equal Opportunity; 41 CFR 60–1.7, Reports and Other Required Information; 41 CFR 60–1.8, Segregated Facilities; 41 CFR 60–250.4, Affirmative Action For Disabled Veterans and Veterans of the Vietnam Era (if in excess of $10,000); and 41 CFR 60–741.4, Affirmative Action for Disabled Workers (if in excess of $2500), wherein the terms "contractor" and "subcontractor" shall mean "Supplier." In addition, orders placed under this Agreement containing a notation that the material or services are intended for use under Government contracts shall be subject to such other Government provisions printed, typed, or written thereon, or on the reverse side thereof, or in attachments thereto.

Identification

Supplier shall not, without Company's prior written consent, engage in advertising, promotion, or publicity related to this Agreement, or make public use of any Identification in any circumstances related to this Agreement. "Identification" means any copy or semblance of any trade name, trademark, service mark, insignia, symbol, logo, or any other product, service, or organization designation, or any specification or drawing of Company or its affiliates, or evidence of inspection by or for any of them. Supplier shall remove or obliterate any Identification prior to any use or disposition of any material rejected or not purchased by Company, and, shall indemnify, defend (at Company's request), and save harmless Company and its affiliates and each of their officers, directors, and employees from and against any losses, damages, claims, demands, suits, liabilities, fines, penalties, and expenses (including reasonable attorneys' fees) arising out of Supplier's failure to so remove or obliterate.

Impleader

Supplier shall not implead or bring an action against Company or its customers or the employees of either based on any claim by any person for personal injury or death to an employee of Company or its customers occurring in the course or scope of employment and that arises out of material or services furnished under this Agreement.

Indemnity

All persons furnished by Supplier shall be considered solely Supplier's employees or agents, and Supplier shall be responsible for payment of all unemployment, social security and other payroll taxes, including contributions when required by law. Supplier agrees to indemnify and save harmless Company, its affiliates, its and their customers and each of their officers, directors, employees, successors and assigns (all hereinafter referred to in this clause as "Company") from and against any losses, damages, claims, demands, suits, liabilities, fines, penalties and expenses (including reasonable attorney's fees) that arise out of or result from: (1) injuries or death to persons or damage to property, including theft, in any way arising out of or occasioned by, caused or alleged to have been caused by or on account of the performance of the Work or services performed by Supplier or persons furnished by Supplier; (2) assertions under Workers' Compensation or similar acts made by persons furnished by Supplier or by any subcontractor or by reason of any injuries to such persons for which Company would be responsible under Workers' Compensation or similar acts if the persons were employed by Company; (3) any failure on the part of Supplier to satisfy all claims for labor, equipment, materials, and other obligations relating directly or indirectly to the performance of the Work; or (4) any failure by Supplier to perform Supplier's obligations under this clause or the INSURANCE clause. Supplier agrees to defend Company, at Company's request, against any such claim, demand, or suit, alleged or proven. Company agrees to notify Supplier within a reasonable time of any written claims or demands against Company for which Supplier is responsible under this clause.

Infringement

Supplier shall indemnify and save harmless Company, its affiliates, its and their customers, and each of their officers, directors, employees, successors and assigns (all hereinafter referred to in this clause as Company) from and against any losses, damages, liabilities, fines, penalties, and expenses (including reasonable attorneys' fees) that arise out of or result from any proved or unproved claim (1) of infringement of any patent, copyright, trademark or trade secret right, or other intellectual property right, private right, or any other proprietary or personal interest, and (2) related by circumstances to the existence of this Agreement or performance under or in contemplation of it (an Infringement Claim). If the Infringement Claim arises solely from Supplier's adherence to Company's written instructions regarding services or tangible or intangible goods provided by Supplier (Items) and if the Items are not (1) commercial items available on the open market or the same as such items, or (2) items of Supplier's designated origin, design, or selection, Company shall indemnify Supplier.

Company or Supplier (at Company's request) shall defend or settle, at its own expense any demand, action, or suit on any Infringement Claim against the other for which it is the indemnitor under the preceding provisions and each shall timely notify the other of any assertion against it of any Infringement Claim and shall cooperate in good faith with the other to facilitate the defense of any such claim.

Inspection

Company's Representatives shall at all times have access to the Work for the purpose of inspection or a Quality Review and Supplier shall provide safe and proper facilities for such purpose.

Insurance

Supplier shall maintain and cause Supplier's subcontractors to maintain during the term of this Agreement: (1) Workers' Compensation insurance as prescribed by the law of the state or nation in which the Work is performed; (2) employer's liability insurance with limits of at least $500,000 for each occurrence; (3) comprehensive automobile liability insurance if the use of motor vehicles is required, with limits of at least $1,000,000 combined single limit for bodily injury and property damage for each occurrence; (4) Commercial General Liability ("CGL") insurance, including Blanket Contractual Liability and Broad Form Property Damage, with limits of at least $1,000,000 combined single limit for bodily injury and property damage for each occurrence; and (5) if the furnishing to Company (by sale or otherwise) of products or material is involved, CGL insurance endorsed to include products liability and completed operations coverage in the amount of $5,000,000 for each occurrence. All CGL and automobile liability insurance shall designate Company, its affiliates, and each of their officers, directors and employees (all hereinafter referred to in this clause as "Company") as an additional insured. All such insurance must be primary and required to respond and pay prior to any other available coverage. Supplier agrees that Supplier, Supplier's insurer(s) and anyone claiming by, through, under or in Supplier's behalf shall have no claim, right of action or right of subrogation against Company and its customers based on any loss or liability insured against under the foregoing insurance. Supplier and Supplier's subcontractors shall furnish prior to the start of Work certificates or adequate proof of the foregoing insurance including, if specifically requested by Company, copies of the endorsements and insurance policies. Company shall be notified in writing at least thirty (30) days prior to cancellation of or any change in the policy.

Invoicing

Supplier's invoices shall be rendered upon completion of the Work or at other times expressly provided for in this Agreement or order; and shall be payable when the Work has been performed to the satisfaction of Company. Supplier shall mail invoices with copies of any supporting documentation required by Company to the address shown on this Agreement or order. The Work shall be delivered free from all claims, liens, and charges whatsoever. Company reserves the right to require, before making payment, proof that all deliverables have been accepted.

Payment Terms

Unless payment terms more favorable to Company appear on Supplier's invoice and Company elects to pay on such terms, invoices shall be paid in accordance with the terms stated in this Agreement, and due dates for payment of invoices shall be computed from the date of receipt of invoice by Company.

Releases Void

Neither party shall require (1) waivers or releases of any personal rights or (2) execution of documents which conflict with the terms of this Agreement, from employees, representatives, or customers of the other in connection with visits to its premises and both parties agree that no such releases, waivers or documents shall be pleaded by them or third persons in any action or proceeding.

Right of Entry and Plant Rules

Each party shall have the right to enter the premises of the other party during normal business hours with respect to the performance of this Agreement, subject to all plant rules and regulations, security regulations and procedures, and U.S. Government clearance requirements if applicable. Supplier shall become acquainted with conditions governing the delivery, receipt, and storage of materials at the site of the Work so that Supplier will not interfere with Company's operations. Storage space will not necessarily be provided adjacent to the site of the Work. Therefore, Supplier shall be expected to select, uncrate, remove, and transport materials from the storage areas provided. Company is not responsible for the safekeeping of Supplier's property on Company premises. Supplier shall not stop, delay, or interfere with Company's work schedule without the prior approval of Company's Representative. Supplier shall provide and maintain sufficient covering and take any other precautions necessary to protect Company's stock, equipment, and other property from damage due to Supplier's performance of the Work.

Supplier's Information

Supplier shall not provide under, or have provided in contemplation of, this Agreement any idea, data, program, technical, business, or other intangible information, however conveyed, or any document, print, tape, disk, semiconductor memory, or other information-conveying tangible article, unless Supplier has the right to do so, and Supplier shall not view any of the foregoing as confidential or proprietary.

Survival of Obligations

The obligations of the parties under this Agreement which by their nature would continue beyond the termination, cancellation, or expiration of this Agreement, including, by way of illustration only and not limitation, those in the clauses *Compliance with Laws, Identification, Impleader, Indemnity, Infringement, Insurance, Releases Void, Use of Information,* and *Warranty,* shall survive termination, cancellation, or expiration of this Agreement.

Taxes

Company shall reimburse Supplier only for the following tax payments with respect to transactions under this Agreement unless Company advises Supplier that an exemption applies: state and local sales and use taxes, as applicable. Taxes payable by Company shall be billed as separate items on Supplier's invoices and shall not be included in Supplier's prices. Company shall have the right to have Supplier contest any such taxes that Company deems improperly levied at Company's expense and subject to Company's direction and control.

Termination

Company may at any time terminate this Agreement or an order, in whole or in part, by written notice to Supplier. In such case, Company's liability shall be limited to payment of the amount due for Work performed up to and including the date of termination (which amount shall be substantiated with proof satisfactory to Company), and no further Work will be rendered by Supplier. Such payment shall constitute a full and complete discharge of Company's obligations. In no event shall Company's liability exceed the price for the Work being terminated.

Tools and Equipment

Unless otherwise specifically provided in this Agreement, Supplier shall provide all labor, tools and equipment (the "tools") for performance of this Agreement. Should Supplier actually use any tools owned or rented by Company or its customer, Supplier acknowledges that Supplier accepts the tools "as is, where is," that neither Company nor its customer have any responsibility for the condition or state of repair of the tools, and that Supplier shall have risk of loss and damage to such tools. Supplier agrees not to remove the tools from Company's or its customer's premises and to return the tools to Company or its customer upon completion of use, or at such earlier time as Company or its customer may request, in the same condition as when received by Supplier, reasonable wear and tear excepted.

Use of Information

Supplier shall view as Company's property any idea, data, program, technical, business, or other intangible information, however conveyed, and any document, print, tape, disk, tool, or other tangible information-conveying or performance-aiding article owned or controlled by Company, and provided to, or acquired by, Supplier under or in contemplation of this Agreement (Information). Supplier shall, at no charge to Company, and as Company directs, destroy or surrender to Company promptly at its request any such article or any copy of such Information. Supplier shall keep Information confidential and use it only in performing under this Agreement and obligate its employees, subcontractors, and others working for it to do so, provided that the foregoing shall not apply to information previously known to Supplier free of obligation, or made public through no fault imputable to Supplier.

Waiver

The failure of either party at any time to enforce any right or remedy available to it under this Agreement or otherwise with respect to any breach or failure by the other party shall not be construed to be a waiver of such right or remedy with respect to any other breach or failure by the other party.

Warranty

Supplier warrants to Company and its customers that material furnished will be new, merchantable, free from defects in design, material, and workmanship, and will conform to and perform in accordance with the specifications, drawings and samples. These warranties extend to the future performance of the material and shall continue for the longer of (1) the warranty period applicable to Company's sales to its customers of the material or of products which incorporate the material, (2) one year after the material is accepted by Company, or (3) such greater period as may be specified elsewhere in this Agreement. Supplier also warrants to Company and its customers that services will be performed in a first class, workmanlike manner. In addition, if material furnished contains one or more manufacturers' warranties, Supplier hereby assigns such warranties to Company and its customers. All warranties shall survive inspection, acceptance and payment. Material or services not meeting the warranties will be, at Company's option, returned for refund, repaired, replaced, or re-performed by Supplier at no cost to Company or its customers and with transportation costs and risk of loss and damage in transit borne by

Supplier. Repaired and replacement material shall be warranted as set forth above in this clause.

Work Done by Others

If any of the Work is dependent on work done by others, Supplier shall inspect and promptly report to Company's Representative any defect that renders such other work unsuitable for Supplier's proper performance. Supplier's silence shall constitute approval of such work as fit and suitable for Supplier's performance.

Mediation

If a dispute arises out of or relates to this Agreement, or its breach, and the parties have not been successful in resolving such dispute through negotiation, the parties agree to attempt to resolve the dispute through mediation by submitting the dispute to a sole mediator selected by the parties or, at any time at the option of a party, to mediation by the American Arbitration Association ("AAA"). Each party shall bear its own expenses and an equal share of the expenses of the mediator and the fees of the AAA. The parties, their representatives, other participants, and the mediator shall hold the existence, content, and result of the mediation in confidence. If such dispute is not resolved by such mediation, the parties shall have the right to resort to any remedies permitted by law. All defenses based on passage of time shall be tolled pending the termination of the mediation. Nothing in this clause shall be construed to preclude any party from seeking injunctive relief in order to protect its rights pending mediation. A request by a party to a court for such injunctive relief shall not be deemed a waiver of the obligation to mediate.

Audit

With the exception of prices fixed by this Agreement, Supplier shall maintain accurate and complete records including a physical inventory, if applicable, of all costs incurred under this Agreement which may affect costs (i.e., travel, living expense, etc.) payable by Company under this Agreement. These records shall be maintained in accordance with recognized commercial accounting practices so they may be readily audited and shall be held until costs have been finally determined under this Agreement and payment or final adjustment of payment, as the case may be, has been made. Supplier shall permit Company or Company's representative to examine and audit these records and all supporting records at all reasonable times. Audits shall be made not later than three (3) calendar year(s) after the final delivery date of material ordered or completion of services rendered or three (3) calendar year(s) after expiration date of this Agreement, whichever comes later.

Authorship, Copyright, and Mask Work Rights

The entire right, title, and interest, including copyright and mask work rights, in all original works of authorship fixed in any tangible medium of expression heretofore or hereafter created by Supplier, or on Supplier's behalf, for Company or furnished to Company hereunder is hereby transferred to and vested in Company. The parties expressly agree to consider as works made for hire those works ordered or commissioned by Company which qualify as such in accordance with the Copyright laws. For all such original works, Supplier agrees to provide documentation satisfactory to Company to assure the conveyance of all such right, title, and interest, including copyright and mask work rights, to Company.

Bankruptcy and Termination for Financial Insecurity

Either party may terminate this Agreement by notice in writing:

1. If the other party makes an assignment for the benefit of creditors (other than solely an assignment of moneys due); or
2. If the other party evidences an inability to pay debts as they become due, unless adequate assurance of such ability to pay is provided within thirty (30) days of such notice.

If a proceeding is commenced under any provision of the United States Bankruptcy Code, voluntary or involuntary, by or against either party, and this Agreement has not been terminated, the non-debtor party may file a request with the bankruptcy court to have the court set a date within sixty (60) days after the commencement of the case, by which the debtor party will assume or reject this Agreement, and the debtor party shall cooperate and take whatever steps necessary to assume or reject the Agreement by such date.

Clause Headings

The headings of the clauses in this Agreement are inserted for convenience only and are not intended to affect the meaning or interpretation of this Agreement.

Developed Information

Supplier agrees that Supplier will and, where applicable, will have Supplier's associates (as defined in the INVENTIONS clause), disclose and furnish promptly to Company any and all technical information, computer or other apparatus programs, specifications, drawings, records, documentation, works of authorship, or other creative works, ideas, knowledge or data, written, oral or otherwise expressed ("Information"), originated or developed by Supplier or by any of Supplier's associates as a result of Work performed under, or in anticipation of, this Agreement. Supplier further agrees that all such Information shall be Company's property, shall be kept in confidence by Supplier and Supplier's associates, shall be used only in performing this Agreement or in the filling of orders hereunder, and may not be used for other purposes except upon such terms as may be agreed upon between the parties in writing. If such Information includes materials previously developed or copyrighted by Supplier and not originated or developed hereunder, Supplier agrees to grant and hereby grants to Company, severally, a nonexclusive, royalty-free license to use and copy such materials. The licenses so granted to Company include the right to grant sublicenses to their subsidiaries and associated companies. Supplier also agrees to acquire from Supplier's associates such assignments, rights, and covenants as to assure that Company shall receive the rights provided for in the DEVELOPED INFORMATION clause.

Harmony

Supplier shall be entirely responsible for all persons furnished by Supplier working in harmony with all others when working on Company's premises or those of Company's customers.

Identification Credentials

Company may, at its discretion, require Supplier's employees to exhibit identification credentials, which Company may issue, in order to gain access to Company's premises for the performance of the Work. If, for any reason, any of Supplier's employees are no longer performing Work, Supplier shall immediately inform Company's Representative in the speediest manner possible. Notification shall be followed by the prompt delivery to Company's Representative of the identification credentials involved or a written statement of the reasons why the identification credentials cannot be returned. Supplier shall be liable for any damage or loss sustained by Company if such identification credentials are not returned to Company.

Inspection and/or Rejection of Work

Supplier shall provide Company with free access to the work performed and the equipment and materials furnished by Supplier under this Contract, if any, for the purpose of inspection thereof. At any time during the progress of the work, Company may condemn or reject any or all of the work, equipment or materials if the same are not in accordance with this Contract and shall give written notice to Supplier of such default. Supplier will thereupon have twenty-four (24) hours to remedy the default. If Supplier fails to timely remedy the default, Company reserves the right to take over any or all work, to provide labor, equipment and materials, and to complete or have completed any part or all of the work. The cost of completion by Company shall be deducted from the unpaid balance, if any, due or which may become due Supplier under this Contract. If there is no unpaid balance or if the cost of completion by Company is in excess of the unpaid balance, Supplier agrees to reimburse Company for such cost, less the amount of the unpaid balance, if any.

Inventions

Supplier agrees that if any inventions, discoveries or improvements are conceived, first reduced to practice, made or developed in anticipation of, in the course of, or as a result of Work done under this Agreement, by Supplier or by one or more of Supplier's employees, consultants, representatives or agents ("associates"), Supplier will assign to Company Supplier's and Supplier's associates' entire right, title and interest in and to such inventions, discoveries and improvements, and any patents that may be granted thereon in any jurisdiction of the world. Supplier also agrees that, without charge to Company, Supplier will and will have Supplier's associates sign all papers and do all acts which may be necessary, desirable or convenient to enable Company at Company's expense to file and prosecute applications for patents on such inventions, discoveries and improvements, and to maintain patents granted thereon. Supplier further agrees to grant and hereby grants Company severally, under any patent issued in any jurisdiction of the world for any invention made prior to the completion of the Work done under this Agreement, nonexclusive, royalty-free licenses (to the extent Supplier has the right to do so) to make, have made, use lease, sell and import any product or facility derived from the Work done under this Agreement. The licenses so granted to Company include the right to grant sublicenses to their subsidiaries and associated companies. Supplier also agrees to acquire from its associates such assignments, rights and covenants as to assure that Company shall receive the rights provided for in this INVENTIONS clause.

Nonexclusive Services

It is expressly understood and agreed that this Agreement neither grants to Supplier an exclusive right or privilege to sell to Company any or all material or services of the type described in this Agreement which Company may require, nor requires the purchase of any material or services from Supplier by Company. It is, therefore, understood that Company may contract with other manufacturers and Suppliers for the procurement of comparable material or services. In addition, Company shall at its sole discretion, decide the extent to which Company will market, advertise, promote, support, or otherwise assist in further offerings of the material or services.

Supplier agrees that purchases by Company under this Agreement shall neither restrict the right of Company to cease purchasing nor require Company to continue any level of such purchases.

Notices

Any notice or demand which under the terms of this Agreement or under any statute must or may be given or made by Supplier or Company shall be in writing and shall be given or made by telegram tested telex, confirmed facsimile, or similar communication, or by certified or registered mail addressed to the respective parties as follows:

> To Company: COMPANY Name
> Address
> Attention:
> To: Supplier
> Supplier address

Such notice or demand shall be deemed to have been given or made when sent by telegram, telex, or facsimile, or other communication or when deposited, postage prepaid in the U.S. mail. The above addresses may be changed at any time by giving prior written notice as above provided.

Representatives

Company's Representative is _____ and Company's Agreement Representative is NAME and Phone Number. All Work rendered under this Agreement is subject to inspection and acceptance by Company's Technical Representative or, in the Technical Representative's absence, by others as may be delegated in writing by Company.

Right of Access

Each party shall permit the other party reasonable access to its facilities in connection with work under this Agreement. No charge shall be made for such visits. It is agreed that prior notification will be given when access is required.

Standards

Employees with records of criminal convictions, other than minor traffic violations, shall not be assigned to Company's premises until a detailed statement of the circumstances is furnished to Company for its review and Company has given its written approval to such assignment. In fulfilling Supplier's obligations under this clause, Supplier shall comply with all laws relating to the making of investigation reports and the disclosure of the information contained therein.

Supplier Employees

The term Supplier employee means anyone performing the Work or furnished by Supplier under this Agreement, including but not limited to the Supplier's employees, consultants, representatives, agents, subcontractors, and subcontractors' subcontractors at all tiers. It is agreed that all persons provided by Supplier to perform the Work are not employees or agents of Company, and Company shall not exercise any direct control or supervision over Supplier employees but Company's Representative will be available for consultation.

Supplier shall be responsible for its own labor relations with any trade or union which represents its employees and shall be responsible for negotiating and adjusting all disputes. Supplier shall be the sole entity responsible for receiving complaints from Supplier employees regarding their assignments and for notifying Supplier employees of the termination or change of their assignments. Company has the right at any time (prior to and after assignment to Company's Work) and for any reason to reject or to have Supplier remove Supplier's employees from the Work under this Agreement upon notice to Supplier. Upon such notice, Supplier shall, at Company's request, replace the Supplier employee(s). In the event of any staffing change, Company shall not be charged for the time required to train the replacement. The amount of non-compensatory training time, if any, shall be mutually determined by Supplier and Company's Representative.

Supplier further agrees that any of Supplier's employees who is or becomes a "leased employee" (as defined in Section 414(n) of the Internal Revenue Code) of Company during the term of this Agreement, shall not be covered by, and shall be excluded from participation in, any employee benefit plan maintained by Company. Supplier shall indemnify and save Company harmless from and against any losses, damages, claims, demands, suits, and liabilities that arise out of, or results from, any failure by Supplier to perform its obligations under this clause. Supplier shall also indemnify and save Company harmless from any entitlement, assertion, or claim, which any of Supplier's employees might have or might make relative to rights or privileges in any Company employee benefit plan and which arises, in whole or in part, out of Work rendered under this Agreement.

Supplier's Information

Except for information owned by Company under the clause DEVELOPED INFORMATION, no specifications, drawings, sketches, models, samples, tools, computer or other apparatus programs, technical or business information or data, written, oral, or otherwise, furnished by Supplier to Company under this Agreement or order, or in contemplation of this Agreement or order shall be considered by Supplier to be confidential or proprietary.

Timely Performance

If Supplier has knowledge that anything prevents or threatens to prevent the timely performance of the Work under this Agreement, Supplier shall immediately notify Company's Representative thereof and include all relevant information concerning the delay or potential delay.

Travel and Living Expenses

In addition to payment of the fee shown in the COMPENSATION clause, reasonable expenses for transportation and living while on approved travel assignments outside the __(define geographic area)__ geographic area previously approved by Company's Representative, shall be reimbursable. Supplier shall submit invoices for reimbursable transportation and living expenses promptly upon completion of the travel events. Supplier shall list the transportation and living charges as separate items on each invoice for the period covered, which shall be in the same detail and accompanied by the same receipts as are required for Company's on-roll employees generally. All invoices shall be certified as approved by Company's Representative prior to submission for payment. Supplier shall retain all such records for a period of not less than one calendar year after the expiration of this Agreement.

Acceptance

When in Supplier's opinion the Technical Services and/or project deliverables described in this Agreement have been completed, Supplier shall provide written notification of such fact to Company.

A. Company shall evaluate each deliverable furnished under this Agreement for compliance with the Specifications and shall submit a written acceptance or rejection to Supplier within thirty (30) days after the receipt by Company of the complete deliverable associated with each task. Such written acceptance or rejection shall be made only by Company's Technical Representative. In no event shall early turnover of the deliverable by Supplier to Company or use of such deliverable by Company or its customers for business, profit, revenue, or any other lawful use constitute acceptance of such deliverable by Company. Company shall have the right to accept portions of any deliverable. Company's acceptance of any deliverable or any portion thereof shall occur only upon a formal written acceptance sent by the aforementioned Company Representative.

B. If a deliverable evaluated pursuant to paragraph A of this clause is rejected, Supplier agrees to correct, at its expense, each error or defect (referred to herein collectively as "defect") leading to such rejection and resubmit the corrected deliverable to Company within fifteen (15) days after receipt of notice from Company of such error or defect. Company shall have thirty (30) days after the resubmitting of such corrected deliverable to accept or reject such deliverable. If the corrected deliverable complies with the Specifications, Supplier shall incorporate the corrections in the deliverable.

C. If the defects in a rejected deliverable are not corrected within the fifteen (15) day period specified in paragraph B of this clause or if a resubmitted deliverable retested or reevaluated by Company during the thirty (30) day re-evaluation period is again rejected, then Company may at its option: (1) retain the deliverable at an equitable adjustment in price as may be agreed by the parties, in which case that deliverable shall be deemed accepted; (2) afford Supplier one or more correction extensions for a period or periods to be specified by Company without prejudice to Company's rights to thereafter exercise its option under either clause (1) or (3) of this paragraph without further notice to Supplier, if the defects have not been corrected; or (3) be entitled to a prompt and full refund of all moneys previously paid under this Agreement. If option (3) is exercised, Company shall have no further obligation to Supplier under this Agreement as to such deliverable or any other deliverable and may elect to terminate this Agreement at any time by written notice to Supplier.

Exhibit 1. Acceptance Form

ACCEPTANCE FORM
CONTRACT TERMS AND CONDITIONS

Clearly state either 1) all Terms and Conditions are acceptable as are, or 2) areas of contention with the Terms and Conditions by referencing the appropriate clause, a brief explanation and alternative language, if any. If a General Agreement exists with Company, please state that fact.

1. I have reviewed the Terms and Conditions and find them acceptable as are as an Agreement for services described in the RFP.

    _____   _____
    Signature   Date

    _____
    Contractor's Name and Title

2. The following Terms and Conditions are not acceptable. In the event an Agreement was to be written for services described in this RFP, the following Clause(s) require negotiation.

    Clause   _____

    _____

    _____

    _____

    Clause   _____

    _____

    _____

    _____

Attach additional page if necessary

NOTE:
Any exceptions, to the following, which are not minor, will disqualify the supplier unless a General Agreement exists:
  • Infringement
  • Indemnity
  • Developed Information, Authorship, Inventions, etc.
  • Warranty

Appendix G

Billing-Related Telecom Concepts and Telecom Glossary

Billing-Related Telecom Concepts

LEC

Term for local telephone company (local exchange carrier)
Assigns a telephone number
Provides dial tone to the subscriber
Enables access to the public switched telephone network (PSTN)

CLEC

Term for competitive local exchange carrier
Provides services that emulate an LEC
Focuses on geographic areas where they can steal customers from an LEC
Some long-distance carriers are also becoming CLECs

Service Providers

Provides conductivity between the customer premise and CLEC or IXC
Can be an LEC where the connection does not go through switching equipment
Also referred to as Bypass
Predominantly for fiber use

PIC

Term for Primary Interexchange Carrier
A three-digit numeric code that assigns a line to a long-distance carrier

PIC for intraLATA calls
PIC for interLATA calls
Pre-subscription for PIC but dial-around possible

Common PICs

AT&T — 288
MCI WorldCom — 222
Frontier Communications — 444
Sprint — 333

Interexchange Carriers (IXCs)

Provider of long-distance services
Access can occur through the PSTN
Lines are PICCed to IXCs

Analog Lines

Single lines — terminated in no-button phones, modem equipment, and fax
  machines
Multi-lines — terminated in multi-button phones, typically using hunting features

Trunks

Single trunk:
        Terminated in PBX equipment for specific purposes
        For example, an executive's private line
Trunk group:
        Contains multiple trunks
        Typically hunts (i.e., finds an available extension out of several in the
          pool)
        Terminated in PBX equipment
DID — Direct Inward Dial
        Telephone company sends three, four, or five digits to the customer
          PBX, which in turns routes call to a specific extension

Centrex Lines

PBX-like features
Intercom
Call forwarding
Call transfer

Digital Connections
Basic Rate ISDN (Integrated Services Digital Network)
        Two B channels and one D channel
        Terminates in a terminal adapter ("digital modem")

Primary Rate ISDN
> 23 B channels and one D channel

Terminates in a channel bank, DSU, or circuit pack within the PBXT1
> 24 voice-grade channels
> Terminates in a circuit pack within the PBX

Attributes of Lines and Trunks

NPA/NXX (uses a structured approach to dialing, including area code and exchange)
Hunting/rotary/terminals
Predefined direction of call flow
Touch-tone
Switched versus dedicated

The North American Numbering Plan

The first six digits of a ten-digit telephone number
The area code and exchange associated with a telephone number

Hunting and Rotary Terminals (telephones)

Multi-line termination allows for a subsequent line to ring if the preceding one is in use or busy
Assignment of line numbers in the hunt groups may not necessarily be in sequential order
Terminal hunting identifies the first number with a unique ten-pilot number, followed by terminal numbers
> Calling party dials a ten-digit number
> Call is routed to the trunk group
> Call can be terminated on any one terminal number (e.g., 0001, 0002)

Predefined Direction of Call Flow

One-way incoming
> Trunk group is designed to handle inbound calls only
> An example use is for 800 termination

One-way outgoing
> Trunk group is designed to handle outbound calls only
> An example use is outbound calls

Two-way
> Trunk group is for either inbound or outbound calls
> Used for overflow when the above two are used

Used stand-alone for economies of scale

Touch-tone

Sometimes an additional cost
Touch-tone is not necessary for inbound services

Switched versus Dedicated

Switched: a single line or quantities of single lines from the local exchange carrier

Dedicated: a T1 (or larger) circuit, with up to 24 voice-grade channels, from the local exchange carrier or the long-distance provider

Toll-Free Calls

A service provided by an IXC for inbound calls toll-free to the originator

Some LEC offer 800 service on an intraLATA basis

800 service terminates or egresses the network

> Switched: directs the call to a ten-digit number over the public switched network
>
> Dedicated: routes the call over facilities that directly connect the IXC with the end user

Toll-free Chargeable Call Features

800/888/877 Numbers

Time and day routing

Caller recognition routing

Announcement features

Area code and exchange routing

Call allocation

Automatic disaster recovery

> Call forward, no answer
>
> Call forward, busy

Trunk group overflow

Time and Day Routing

Uses geographically diverse locations to cover after-hours calls

When an East Coast location closes at 6 p.m., calls are routed to the West Coast facility, which stays open until 12 midnight, for nine additional hours of operation

When the East Coast location opens at 7 a.m., it can answer calls from the West Coast, where the time is 4 a.m.

Caller Recognition Routing

Automatically routes customers based on unique needs

Premium customers are identified by originating number and routed to specifically assigned representatives

Announcement Features

Professionally recorded greetings and announcements

Example: Auto Attendant in the sky:
   "Press 1 for Sales, 2 for Marketing, 3 for Sales"
   Routes to different physical locations

Area Code and Exchange Routing

Routes calls based on the originating number
For example, callers who dial 1-800-pizza2go
   Routes caller to the nearest "pizza" eatery
   Callers from New York would be routed to (212) 555-PIZZ
   Callers from California would be routed to (415) 555-SPAZ

Call Allocation

Distributes calls, by percentage, to specific locations based on availability of
   personnel
Example:
   Two call centers have a 40/60 staffing distribution
   Calls are routed based upon this distribution

Automatic Disaster Recovery

If an initial point of destination for an inbound call is not available, call is
   automatically rerouted
There are two types of automatic disaster recovery:
   Call forward, no answer: if the call is not answered in a predetermined
      period, it is directed to an alternate location
   Call forward, busy: if all service to the original destination is in use, the
      call is directed to an alternative location

Trunk Group Overflow

800 numbers rerouted to a second designated trunk group when the initial
   trunk group is not available
Difference between automatic disaster recovery (ADR):
   ADR refers to calls routed to an alternate location
   Trunk group overflow refers to the same location, but a different trunk
      group

Telephone Equipment

Single-line phones
Multi-line phone systems
PBX — private branch exchange
Hybrid

Single-Line and Multi-Line Phones

Single-line phones
      No lights, no hold
      2500 set, touch-tone deskset

Multi-line phones
    Lights
    Hold
    Multi-line appearances
    Avaya product: Partner

Private Branch Exchange (PBX)

Supports many users, and many types of phone styles
Shared use of local and long-distance lines provides economies of scale
Facilitates internal communications by intercom-like functions
Avaya product: Definity
Nortel product: Meridian

Hybrid System

Connected to the network
    Lines or trunks
    T1
Provides limited PBX features
Provides more features than a key system
Avaya product: Merlin Legend
Nortel product: Norstar

General Telecom Glossary

Access Charges — A charge by the local telephone company, for use of the local company's exchange facilities, and/or interconnection with the telecommunications network. Long-distance companies rely on the loops, switches, and transport facilities of local telephone companies for access to their customers. As a result, local telephone companies recover a portion of their costs from long-distance companies accessing their networks. Both the manner in which these access charges have been assessed and the proportion of the costs they have recovered have varied considerably over time.

Access Control List (ACL) — Used to control access to an IP network. Access lists are typically found on routers. A group of statements on the router defines a pattern that would be found in an IP packet. As packets are encountered, they are matched with a control list and accepted or rejected as appropriate.

Access Line — The facilities required to provide access to the local and switched network. These facilities are positioned between a serving central office and the customer. The access line includes the non-traffic-sensitive central office equipment, the subscriber loop, the drop line, inside wiring, and the main jack.

Alternate Mark Inversion (AMI) — A line coding where 0 volts represents 1s; alternate positive and negative volts represent zeros.

Analog — Generally represents the concept of a smooth and continuous variation in electrical signal or amplitude. In contrast, digital communications are characterized by fixed states such as 0 and 1, off and on, and positive or negative voltage.

Application Service Provider (ASP) — ISPs doing e-mail and Web hosting today are the pioneers of the ASP market. Many mission-critical applications are being offered by ASPs, including financial management systems. A company may choose an ASP for reasons such speed to market, cost, and timing. The ASP takes responsibility for hardware and facilities management. The applications are generally owned or managed by the ASP supplier where pricing is usually per transaction and user seat.

Asymmetric Digital Subscriber Line (ADSL) — One of a variety of DSL offerings for bandwidth-heavy applications such as the Internet, broadcast video, and distance learning. Data rates from 64 kbps to 8.192 Mbps on the downstream and from 16 to 768 kbps on the upstream channel. It can be splitter or splitterless with a pair of ADSL modems installed at the customer site and one point of termination in the network.

Asynchronous Transfer Mode (ATM) — ATM is an International Telecommunications Union standard for cell relay. Information is conveyed in small, fixed-size cells. It is capable of very high speeds and is used for the transport of voice, video, data, and images.

Automatic Call Distributor (ACD) — Software and hardware that intercepts calls coming into a telephone switch and directs the call to particular agents or help desk personnel based on predefined criteria.

Automatic Number Identification (ANI) — In the residential world, this functionality is referred to as "caller ID." Header information on the call provides the calling number in a standard format that can be displayed and read by telephony systems.

Automatic Route Selection (ARS) — Electronic or mechanical selection and routing of outgoing calls without human intervention. Typically, ARS is used to reduce the cost of outgoing calls by using preprogrammed information to determine the least-cost route. For example, an organization may have a tie line from Houston, Texas, to Brighton, Michigan, as well as the usual PSTN connections. Physically, an outgoing call could go either via the dedicated tie line or via the PSTN. Because the tie line is an already-paid-for resource, ARS would send the call through the tie line rather than over the PSTN.

Bandwidth — The width of a communications channel. Bandwidth is either measured in Hertz (Hz, cycles per second) for analog communications or kilobits per second (kbps) for digital communications.

Bell Operating Company (BOC) — One of 22 regulated local telephone companies of the former Bell system. Organized into seven Regional Bell Operating Companies (RBOCs), including Southern Bell & Telegraph Company, New York Telephone, The Ohio Bell Telephone Company, and The Pacific Bell Telephone and Telegraph Company.

Billing Telephone Number (BTN) — Multiple WTNs (working telephone numbers) get assigned to one BTN (essentially the account number) for billing purposes; a BTN can also be a WTN, but a WTN is not necessarily a BTN; one BTN is usually assigned per location.

Broadband — A general term indicating WAN transmission speeds above 1 megabit per second (1 Mbps). Broadband includes services such as cable modem, DSL, wideband wireless Internet connections, digital broadcast-delivered enhanced services, various streaming media offerings, and two-way satellite Internet connectivity. *Note:* Some consider broadband to be T3 speeds and above.

Bypass Facilities — These serve as alternatives to local telephone company provided distribution. Facilities such as cellular radio, two-way cable TV, short-haul microwave, and direct satellite to rooftop antennas are examples of technology employed by other Common Carriers (OCCs) and cable TV companies that have the capability to circumvent the use of the local phone network.

Carrier ID Code (CIC) — Synonymous with a PIC.

Cellular Telecommunications & Internet Association (CTIA) — An international organization that represents all elements of wireless communication — cellular, personal communication services, enhanced specialized mobile radio, and mobile satellite services. It serves the interests of manufacturers, service providers, academics, and others.

Central Exchange (Centrex) — Provides PBX features and telephone sets without having to purchase the PBX itself; service is provided by LECs; prices for services tend to be competitive; services are usually rented on a line-by-line basis. Main components are the serving CO (Central Office); the CPU; the station, line, and power cards; and the switch.

Central Office (CO) — Telephone company building where subscribers' lines are connected to switching equipment (usually within two to three miles of home or office).

Channel Associated Signaling (CAS) — One of two signaling types used by T1s.

Channel Service Unit (CSU) — Interface device between the telephone company digital line and the customer's equipment in the United States.

Code Division Multiple Access (CDMA) — CDMA is a digital cellular technology that uses spread-spectrum techniques. This protocol does not assign a specific frequency to each user. Rather, each channel uses the full available spectrum. A pseudo-random digital sequence is used to code specific conversations, resulting in greatly reduced interference.

Code Exciter Linear Predictor (CELP) — An encoder that compares speech samples with an analytical model of the voice, and computes the errors between the speech and the model. It can produce very high-quality speech

Coder/Decoder (Codec) — Converts analog signals to digital. For example, a cellular phone has a codec that converts the analog spoken word (translated into electrical signals) into 1's and 0's that characterize digital information.

Committed Information Rate (CIR) — For Frame Relay services, CIR guarantees a specified amount of bandwidth (measured in kilobits per second). Typically, when purchasing a Frame Relay service, a firm will ask for a specific CIR. The Frame Relay network vendor guarantees that bandwidth not exceeding this quantity of traffic will be delivered. More traffic may be carried by the circuit, but it is not guaranteed.

Common Carrier — A supplier in an industry that undertakes to "carry" goods, services, or people from one point to another for the public in general or for specified classes of the public. In telecommunications, such "carriage" relates to provision of transmission capability over the telecommunications network. A communications common carrier is subject to federal and state regulatory commissions.

Common Channel Interoffice Signaling (CCIS) — An electronic means of signaling between any two switching systems independent of the voice path. The use of CCIS makes possible new services, versatile network features, more flexible call routing, and faster call connections.

Common Channel Signaling (CCS) — One of two signaling types used by T1, CCS allows the D channel to be shared between several T1s and therefore enables slightly more throughput (24 channels versus 23) for those T1s that rely on the common channel. In the authors' opinion, using a D channel per T1, while slightly wasteful, is less risky. If the "common" channel does down, under CCS many T1s would fail, not just one.

Common Control Switching Arrangements (CCSA) — A service by carriers that furnishes an inter-city private-line switching network for a customer.

Community Antenna Television/Cable TV (CATV) — A broadband communications technology in which multiple television channels, as well as audio and data signals, may be transmitted either one way or bidirectionally through an often hybrid (fiber and coaxial) distribution system to a single or to multiple specific locations. CATV originated in areas where good reception of direct broadcast TV was not possible. Now CATV also consists of a cable distribution system to large metropolitan areas in competition with direct broadcasting. The abbreviation CATV originally meant community antenna television. However, CATV is now usually understood to mean cable TV.

Companding — An operation in which the dynamic range of signals is compressed before transmission and is expanded to the original value at the receiver. Note: The use of companding allows signals with a large dynamic range to be transmitted over facilities that have a smaller dynamic range capability. Companding reduces the noise and crosstalk levels at the receiver.

Competitive Local Exchange Carrier (CLEC) — Competitor for local, long-distance, international, Internet access, and entertainment services that uses its own switches or resells ILEC services. Includes ISPs, IXCs (AT&T, Sprint, etc.) and CATV providers.

Computer Telephony Integration (CTI) — The set of software, hardware, and standards that allows computers and telephony systems to interact so that the power of the computer can be used in conjunction with the voice system. For example, if a customer calls a call center, the computer can use ANI to look up information in a database and "pop" relevant information onto a workstation screen (e.g., the customer's buying and credit history).

Convergence — The integration of various technologies into a single technical base. One example of convergence is the melding of voice, data, and video into a packet-based architecture riding on the same infrastructure. Another example is the merger of TV set-tops and PCs.

Cramming — The unethical (sometimes illegal) placing of unwanted/unneeded services on a telephone bill. For example, voicemail might be added to an emergency Centrex line.

Customer Premises Equipment (CPE) — All terminal equipment located on the customer premise, both state and interstate, except coin-operated telephones, and encompassing everything from telephones to the most advanced data terminals and PBXs. Term is derived from FCC Second Computer Inquiry decision.

Customer Service Record (CSR) — A key report from the local telephone company that shows what services are in place and their costs. Much of this information does not appear on the monthly bill. CSRs must be specifically requested.

Data Over Cable Service Interface Specifications (DOCSIS) — DOCSIS is a standard interface for cable modems; it handles incoming and outgoing data signals between a cable TV operator and a computer or television set. Cable communications are "shared," meaning that all nodes on the network can see each other's traffic (at least theoretically).

Data-Link Connection Identifier (DLCI) — A unique number that tells Frame Relay how to route the data. The DLCI field identifies which logical circuit the data travels over.

Dedicated Access — Direct access to IXC without the assistance of an LEC; in this situation, the IXC does not have to pay the LEC charge.

Digital Access Cross-Connect System (DACS) — A digital switching device in telecommunications for routing T1 lines. The DACS can cross-connect any T1 line in the system with any other T1 line also in the system (think of a cat's cradle). DACS can go even further, connecting any DS-0 channel on a T1 line to any DS-0 time slots of any other line.

Digital Loop Carrier (DLC) — DLC allows carriers to run a four-wire, twisted pair circuit from the central office to a remote location. Prior to that, individual copper wires were proliferating at an alarming rate. DLC became popular in the mid-1970s and increased the carrying capacity (channels) of the network. This system is sometimes called "pair gain."

Digital Private Network Signaling System (DPNSS) — A private networking standard, DPNSS is a standard developed by the manufacturers of large telephony equipment to allow interconnection between their equipment over the ISDN network.

Digital Signal (Level 0) — (DS-0) — 64 kbps

Digital Signal (Level 1) — (DS-1) — 1.544 Mbps

Digital Signal (Level 2) — (DS-2) — Not used in the industry

Digital Signal (Level 3) — (DS-3) — 45 Mbps

Digital Subscriber Line (DSL) — Copper wire carrying voice communications have historically been throttled to a specific frequency range. By introducing new protocols and special hardware attached to both the user and switch ends of the line, data can be transmitted over the wires at far greater speed than with the standard phone wiring. The distance limitation is around 12,000 feet from the central office. However, new technologies are extending this distance.

Digital Subscriber Line Access Multiplexer (DSLAM) — DSLAM terminates multiple DSL lines and aggregates traffic from them at the CO.

Digital Transmission System (DTS) — A wideband local distribution system that provides two-way transmission of high-speed digital electronic message services. DTS permits end-to-end microwave-based intra-city links of digital communications networks. Using inter-city terrestrial microwave and satellite links, DTS typically employs transmitting and receiving stations throughout a metropolitan area, with transceivers located at a subscriber's premises.

Direct Inward Dialing (DID) — With direct inward dialing, a caller dials a number, which is then sent via the telephone company (or privately, if a network is in place) to specific trunks that terminate at the destination organization. The PBX on premises picks up the dialed number (sometimes termed DNIS) and routes it to the appropriate extension. Without DID, callers would reach the central operators and have to be routed manually to the desired extension.

Drop, subscriber's — Wire that runs from a cable terminal or an open wire bridging point to the subscriber's house.

E1 — Similar to a T1 but with 30 channels; used in Europe.

Echo Return Loss (ERL) — Difference in decibel level of outgoing speech and returning echo.

Electronic Switching System (ESS) — First digital switch; developed by AT&T.

Ethernet — Ethernet is the most widely installed local area network (LAN) technology. Specified in a standard, IEEE 802.3, Ethernet is a Layer 2 protocol. LANs using Ethernet are typically linked via coaxial cable, twisted pair (e.g., CAT5e or CAT7) or wireless links. The most commonly installed Ethernet systems are called 10BASE-T and provide transmission speeds up to 10 Mbps. Much higher speeds are available, such as 100 Mbps and Giganet Ethernet.

Exchange — Defined as a geographic area where telephone services and prices are uniform; one or more central offices (COs) serve an exchange area. All calls made within any two points in an exchange area are classified as local calls. Revenue generated from local calls belongs to the LEC.

Exchange Service — The furnishing of ordinary voice-grade telecommunications service under regulation within a specified geographic area.

Extensible Mark-Up Language (XML) — Includes many sub-protocols such as VoiceXML, which allows standard voice interactions between the browser and the user.

Facsimile (Fax) — Fax technology includes a scanner for converting an image to digital bits, a digital signal processor for eliminating white space, and a modem for converting bits into an analog signal. Virtually all faxes now communicate at the Group 3 standard (Group 3 enhanced and other, higher standards exist but all modern faxes will fall back to basic Group 3 as needed).

Federal Communications Commission (FCC) — The federal agency empowered by law to regulate all interstate and foreign radio and wire communications services originating in the United States, including radio, television, facsimile, telegraph and telephone systems. The agency was established under the Communications Act of 1934.

Fiber Distributed Data Interface (FDDI) — FDDI is a multi-vendor LAN standard developed by the American National Standards Institute (ANSI). It is intended for organizations that need high-speed LAN bandwidth. FDDI is similar, in some ways, to Token Ring. Because it is very difficult to tap into a fiber, a ring was the logical solution. FDDI groups stations — including workstations, bridges, and routers — into a ring, with each station having an input fiber from the previous station and an output to the following one. The last station connects back to the first to complete the ring.

Fiber Optics — The technology of guiding and projecting light for use as a communications medium. Hair-thin glass fibers that allow light beams to bend and reflect with low levels of loss and interference are known as "glass optical waveguides" or simply as "optical fibers."

Fiber to the Curb (FTTC) — Refers to the installation and use of optical fiber cable directly to the curbs near homes or any business environment.

Fiber to the Home (FTTH) — Use of an all-fiber network to bring broadband directly to the home. Fiber can theoretically provide vastly more bandwidth than either DSL or cable modem.

File Transfer Protocol (FTP) — The protocol used on the Internet for sending files.

Foreign Exchange (FX) — Telephone exchange service furnished to a customer through a central office of an exchange other than the exchange in which the customer is located.

Fractional T1 (FT-1) — Fractional T1s use the same technology and connectivity as full point-to-point T1s. However, with this service, a full 1.54-Mbps T1 is split up to cater to an organization's bandwidth needs.

Frame Relay (FR) — Successor to the X.25 Protocol, Frame Relay is a form of packet switching that can accommodate data packets of variable length and does not do higher level protocol conversion. It is efficient and relatively inexpensive.

Frame Relay Assembler/Disassembler (FRAD) — A communications device that breaks a data stream into frames for transmission over a Frame Relay network and recreates a data stream from incoming frames. Stand-alone FRADs are less commonly used now because a router configured for Frame Relay can accomplish the same task and better handle congestion.

Frequency Division Multiplexing (FDM) — Divides the frequency spectrum among logical channels, with each user having exclusive possession of a specific frequency band (compare to TDM); requires analog circuitry.

Gigabits per Second (Gbps) — Gbps represents sufficient bandwidth to accommodate video easily. However, it pales in comparison with what is coming. On March 22, 2002, scientists at Bell Labs announced they had transmitted 2.56 terabits (trillion bits) of data across a distance of 4000 kilometers.

Global Positioning System (GPS) — A worldwide radio-navigation system. It is supported by 24 satellites and their associated ground stations. GPS uses these satellites as reference points to calculate positions accurate to a matter of meters.

Global System for Mobile Communications (GSM) — GSM is a digital cellular phone service that is heavily used in Europe, Japan, and other countries — and now increasingly in the United States. Currently, more than one in ten of the world's population uses GSM. This standard (actually a collection of standards) has a migration path to third-generation broadband capability for data, in addition to its existing voice services. Originally Groupe Speciale Mobile.

High-Bit Rate DSL (HDSL) — HDSL is a high bit-rate Digital Subscriber Line. It can deliver T1 speeds. At present, this speed requires two lines.

High-Level Data Link Control (HDLC) — HDLC corresponds to Layer 2 (the data link layer) of the ISO seven-layered architecture. It is responsible for the error-free movement of data between network nodes.

Hypertext Markup Language (HTML) — HTML is used for publishing hypertext on the World Wide Web. It is a nonproprietary format based on SGML. It can be created and processed by tools from simple editors such as WordPad, to sophisticated WYSIWYG authoring tools such as Front Page or GoLive.

Hypertext Transfer Protocol (HTTP) — HTTP is an application-level protocol that supports distributed, collaborative, hypermedia information systems. It is an object-oriented protocol that can be used for functions such as name servers and distributed object management systems. HTTP handles data representations, freeing application developers to work with systems independent of the data being transferred.

Incumbent Local Exchange Carrier (ILEC) — Name given to an LEC that provided local phone service prior to the Telecommunications Act of 1996.

Independent Carriers — All telephone companies that are not affiliated with BOCs. There are over 1400 independents in the United States.

Integrated Analog Network (IAN) — Network where all switches, interoffice trunks, local loops, and telephones are analog.

Integrated Digital Network (IDN) — Network where all switches, interoffice trunks, local loops, and telephones are digital.

Integrated Digital Subscriber Line (IDSL) — IDSL is DSL at a specific rate — 144 kbps. IDSL uses ISDN transmission coding, bundling together both ISDN channels and voice all on one circuit. For some users who are too far from the CO to use standard DSL, IDSL is the best option, although it is somewhat slower than most DSL services.

Integrated Services Digital Network (ISDN) — ISDN is a system of digital phone connections that has been available for many years. It was once considered the fair-haired child of telecom but is now viewed as a rather mundane product superseded by DSL, cable modem, and other high-bandwidth technologies. It is, however, digital and in some locations is the best wireline link available below standard (and more expensive) T1 connections. With ISDN, voice and data are carried by bearer channels (B channels) occupying a bandwidth of 64 kbps. A data channel (D channel) handles signaling at 16 kbps or 64 kbps, depending on the service type. Dial-up ISDN is often used as a digital failover circuit, brought into service when the primary line (e.g., Frame Relay) is down.

Interactive Voice Response (IVR) — IVR is a telephony system in which an individual uses a touch-tone telephone to extract or supply information from/ to a database. IVR systems are the familiar "Press 1 for your account balance, Press 2 to hear your last ten transactions…" applications. IVR technology is used for surveys, pricing lookup, and a multitude of other applications. IVR differs from CTI in that CTI has as its major input the workstation keyboard/ monitor, whereas IVR uses the touch-tone telephone keypad.

Interexchange Carrier (IXC) — A long-distance company. IXCs carry interstate, interLATA, and international traffic. Examples include MCI WorldCom, Sprint, and AT&T long-distance services.

InterLATA — Across one or more LATAs. A call from one LATA to another LATA, usually handled by an IXC.

International Record Carrier (IRC) — Carrier providing overseas/international telecommunications services, other than voice telecommunications (e.g., tele- typewriter, facsimile, and data)

International Telecommunications Union (ITU) — According to the ITU itself, "The Union was established last century as an impartial, international organi- zation within which governments and the private sector could work together to coordinate the operation of telecommunication networks and services and advance the development of communications technology. Whilst the organiza- tion remains relatively unknown to the general public, ITU's work over more than one hundred years has helped create a global communications network which now integrates a huge range of technologies, yet remains one of the most reliable man-made systems ever developed."

Internet Protocol (IP) — IP is the protocol (along with TCP) upon which the Internet and much packet-based communications is based. IP is a Layer 3 protocol.

Internet Service Provider (ISP) — A company that provides Internet access. Examples include AOL, UUNET, MindSpring, etc.

Interoffice call — Local call placed by a customer of one CO to a customer served by a different CO within the same exchange area. The CO initiating the call must switch the call to another CO.

IntraLATA — Within a LATA; a call that begins and ends within the same LATA

Intra-state — Within a state. Intra-state is not the same as intra-LATA. Calls within the same state can cross LATA boundaries.

Kilobits per Second (kbps) — kbps is more applicable to the dial-up world (which tops out around 53 kbps, not 56 kbps). Broadband is usually stated in megabits per second (Mbps).

Lightweight Directory Access Protocol (LDAP) — LDAP is a specification for a client/server protocol to retrieve and manage directory information. Originally intended as a method for clients on workstations to access X.500 directories, its use has expanded to many other applications. LDAP is increasingly accepted as the standard protocol/system to house central directory information (includ- ing telephony, LAN, and organizational information).

Local Access and Transport Area (LATA) — Geographic service areas (exchange) of "common social, economic, and other purposes" established by the Bell Operating telephone companies. An exchange can be serviced by one or more central offices furnishing ordinary voice-grade service. There are 210 LATAs in the 48 states and Washington, D.C.

Local Area Network (LAN) — A LAN connects personal computers, servers, printers, and other devices over a relatively short distance (e.g., inside a building). LANs can be wireline (with CAT7 cabling, for example) or wireless. With special equipment and protocols, LANs can be extended to cover larger geographic areas.

Local Exchange Carrier (LEC) — Telephone company that owns and operates the switching equipment within a specific geographic area. Handles intraLATA traffic (i.e., within a LATA). Includes Southwestern Bell (SBC), Quest and Verizon.

Local Loop — Uses a single wire pair to carry a signal from the CO to a business or residence. It is optimized to carry human voice.

Local Multipoint Distribution Service (LMDS) — LMDS is a broadband, two-way radio frequency service in the 27- to 31-GHz band. It is used to transmit MPEG-2 quality video and is considered a wireless alternative to fiber and coax.

Local Number Portability (LNP) — Allows telephone customers in the United States to keep their local phone number if they switch to another service provider. LNP is required of local carriers as the cost of being allowed to compete in the long-distance market.

Media Access Control (MAC) — On a network, the MAC address is the computer's unique hardware number. Other devices on the network, such as servers, also have a MAC address. On an Ethernet LAN, it is the same as the Ethernet address. When connected to the Internet from a workstation, a correspondence table relates the IP address to the computer's physical (MAC) address on the LAN.

Megabits per second (Mbps) — Millions of bits per second. Many in the industry consider one megabit per second to be the starting point for broadband. Others consider broadband to start at 45 Mbps (T3 speed).

Metropolitan Area Network (MAN) — Considered a "LAN on steroids," the MAN links users in a region larger than a LAN but smaller than a WAN. A MAN typically connects multiple smaller networks in a city. Sometimes a MAN is considered to be the interconnection of several LANs via a backbone.

Microwave Multipoint Distribution System (MMDS) — MMDS is a broadband wireless technology that delivers voice, data, Internet, and video services in the 25-GHz and higher spectrum (it varies, depending on licensing).

Mobile Identification Number (MIN) — A digital representation of the ten-digit directory number assigned to a mobile station. This is the telephone number stored in the Subscriber Unit for North American cellular systems.

Motion Picture Experts Group (MPEG) — MPEG combines JPEG to compress individual images and create a form of delta-values where two adjacent frames are compared and only the difference is transmitted. MPEG-1 is similar in quality to a standard videotape playback and is also the basis for the MP3 audio format; MPEG-2 is higher-quality video.

Multiplexing — Simultaneous transmission of multiple messages in one channel over a telephone network.

Multi-Protocol Label Switching (MPLS) — MPLS speeds up traffic and makes the network easier to manage. The protocol sets up a defined path for a given sequence of packets. The path is identified by a label in each packet and therefore the router is saved the time needed to look up the address of the next node to forward the packet to. It works with IP, ATM, and Frame Relay.

Network Interface Card (NIC) — The NIC is an expansion board inserted into a computer so the computer can be connected to a network. Typically, NICs are specific to a type of network, protocol, and media.

Network Operations Center (NOC) — A telecom network is monitored and managed from a NOC. Network management software (such as Tivoli and HP Openview) and hardware is used to track the network and maintain operations. Many NOCs are 7x24 and have a plethora of reports and displays that help manage events.

North American Numbering Plan (NANP) — A plan for the allocation of unique ten-digit address numbers. The numbers consists of a three-digit area (numbering plan area) code, an office code, and a four-digit line number. The plan also extends to format variations (e.g., three- and seven-digit address), prefixes (e.g., 1, 0, 01, and 011), and special code applications (e.g., service access codes).

One Flat Rate (1FR) — Residential line for ordinary telephone service. This is in contrast to a measured service.

One Measured Business (1MB) — Analog business line. Sometimes this line is replaced by Centrex service.

Operations Support Systems (OSS) — OSS includes the systems that perform inventory, engineering, planning, management, and repair services for carriers and their networks.

Optical Carrier - Level 3 (OC-3) — A protocol designed to operate over fiber at a speed of 155.52 Mbps.

Packet Switching — A data communications switching and transmission system whereby an input data stream is broken into uniform data "packets" to which is appended addressing information, sequence counts, and error controls. Each packet is transmitted independently through the network so as to maximize the utilization of transmission facilities. At the receiving end, the packets are checked for errors, resequenced as necessary, and combined into an output data stream.

Permanent Virtual Circuit (PVC) — A permanent virtual circuit (PVC) is a software-defined logical connection in a network such as a Frame Relay network. A feature of Frame Relay that makes it a highly flexible network technology is that users (companies or clients of network providers) can define logical connections and required bandwidth between end points and let the Frame Relay network technology worry about how the physical network is used to achieve the defined connections and manage the traffic.

PIC-Freeze — With a PIC-freeze in place, only a specific, authorized party can change an organization's LD carrier with LEC. A PIC-freeze prevents unauthorized "slamming." By requiring a written statement on letterhead with a preauthorized signature, unscrupulous parties find it difficult to make unauthorized changes.

Plain Old Telephone Service (POTS) — This is the service shown on old 1930s films — the standard telephone service that is still used in most homes. Analog signals are carried on copper wires and bandwidth is limited to around 52 kbps. Newer, high-speed transmission services, such as IDSN and xDSL, are not POTS, although they can use the same copper going to the home or business.

Point of Presence (POP) — A switching office built by an IXC to allow the IXC to handle calls originating in a LATA.

Point-to-Point (PTP) — A dedicated link that connects only two stations. A point-to-point circuit is often a T1 link but could be any type of connection whose entire bandwidth is dedicated to transmitting data, voice, or video between two locations.

Post, Telegraph, and Telephone (PTT) — Governmental telecom organizations outside the United States that operate their respective country's communications functions.

Primary Interexchange Carrier (PIC) — Three-digit numeric code that assigns a line to a long-distance carrier.

Primary Interexchange Carrier Charges (PICC) — The PICC is a flat, per-line charge assessed by the local exchange carrier (LEC) on a customer's presubscribed interexchange carrier. In the case of an end user who has not preselected a specific IXC, the LEC is permitted to recover the charge directly from that end user.

Primary Rate Interface (PRI) — As part of the ISDN specification, PRI channels are carried on a T-carrier system line (in the United States, Canada, and Japan) or an E-carrier line (in other countries). It is important to note that PRIs are T1s but not all T1s are PRIs. With the 23 bearer (B) channels and one D (management) channel, PRIs are quite flexible; channels can be designated for different uses (data, voice, video) and also can be designated as inbound and outbound at the same time. The older T1 specification requires that channels be designated as either inbound or outbound, resulting in some inefficiency. PRI circuits are linked directly to the CO.

Private Automatic Branch Exchange (PABX) — Alternate terminology for the PBX. Used chiefly outside the US.

Private Branch Exchange (PBX) — A PBX is a telephone switch that resides in an organization other than the telephone company's central office. The PBX allows individuals within the building or plant to call each other without going through the public network (hence fewer digits need be dialed). In addition, a one for one relationship between user telephones and lines to the CO need not exist. Because the PBX can "switch" calls, trunks can be reused so that various people can be reached using the same incoming lines. Many employees can likewise use the same outgoing trunks. Key systems, on the other hand, have traditionally been structured so that a line is associated with a "key" on a telephone (e.g., "Artie, pick up on line 2.").

Private Line Service (PLS) — Initially, private line service was point-to-point telecommunications service over a channel dedicated to a particular customer's private use. FCC regulation now allows parts of private line services, except access lines, to be used in common by many customers. Private line services are used by customers with high volume or specialized requirements. However, most private lines are connected directly or indirectly (e.g., via a PBX to the public switched network). Another definition of Private Line Service is an outside (Centrex-like) telephone number separate from the PBX. Sometimes, these are used as emergency phones, should the PBX fail.

Public Key Infrastructure (PKI) — PKI is a standard technique for authenticating a message sender or encrypting a message on the Internet. Less robust cryptography (symmetric cryptography) requires a secret key to be shared between two parties and therein lies its weakness; during transit, the key could be discovered by an unauthorized party. PKI does not require sharing of the complete key.

Public Switched Telephone Network (PSTN) — A domestic telecommunications network usually accessed by telephones, key telephone systems, private branch exchange trunks, and data arrangements. Completion of the circuit between the call originator and call receiver in a PSTN requires network signaling in the form of dial pulses or multifrequency (DTMF) tones. The PSTN started when Alexander Graham Bell reportedly shouted, "Watson, come here" into the first primitive telephone. We will never know what Bell actually said, but because, due to an accident, acid was eating through his clothes, the words may not have been quite that genteel.

Public Utility Commission (PUC) — The following definition, from the Public Utility of Texas, provides a sense of the public interest oversight of telecommunications provided by PUCs: "The Public Utility Commission of Texas (PUC of Texas) supports the efforts of the Federal Trade Commission (FTC) to prevent cramming and strengthen the 900 rule. Customers frequently assume their local telephone bill still contains only legitimate, regulated charges, but with recent changes in the marketplace unregulated and unauthorized charges now appear on local phone bills. The telephone bill is used as a collection vehicle, similar to a credit card but without the security, dispute resolution procedures, or authorization requirements that accompany credit card transactions. As a result, the local telephone bill is sometimes used to commit fraud."

Pulse Code Modulation (PCM) — Process of converting analog sample to digital bitstream; performed by a codec (coder/decoder). It is the basis of the modern telephone system.

Regional Bell Holding Company (RBHC) — Same as RBOC or BOC.

Regional Bell Operating Company (RBOC) — Same as RBHC or BOC. An RBOC is a LEC, but a LEC is not necessarily an RBOC.

Remote Access Server (RAS) — A server that provides remote (not local) access to authorized users. Using dial-up, ISDN links, VPN, or some other means of access, users can gain access to files and other resources that are located remotely. Aside from bandwidth considerations, it is as if the user was on site, although he or she is at a hotel or working from home. Robust authentication in the RAS is essential.

Remote Call Forwarding (RCF) — Automatically forwards a call from an old number to a new one, or from one number to another number chosen in advance. Remote call forwarding should be reviewed carefully because it has been frequently used by "phone phreaks" to perpetrate toll fraud. If a hacker can forward a phone in an office building to an international location, then anyone can dial locally to that primary number and be forwarded — at the organization's expense — to the international location.

Service Level Agreement (SLA) — A written agreement between a consumer of services and a provider of services. The agreement specifies quantitative and qualitative levels of service as well as procedures to be followed (e.g., escalation procedures if lines are down).

Signaling System 7 (SS7) — SS7 is an entire architecture, using out-of-band signaling to support the PSTN (public switched telephone network). It supports call setup, billing, routing, and telecom database exchange functions of the PSTN. For example, SS7 allows toll-free numbers to be dialed by coordinating databases that keep up with the terminating number as well as the entity that will pay for the call.

Simple Mail Transfer Protocol (SMTP) — SMTP is part of the TCP/IP protocol suite. Obviously used for sending and receiving e-mail, it is often accompanied by two other protocols, POP3 or IMAP, to improve its ability to queue messages at the receiving end.

Simple Network Management Protocol (SNMP) — SNNP is a set of protocols for managing complex networks. SNMP sends messages, called protocol data units (PDUs), to different parts of a network. Related software, called MIBs (management information databases), stores data about SNMP-compliant devices (e.g., routers). SNMP can read these databases and return status information to anyone requesting it.

Small Office/Home Office (SOHO) — Generic term for a small networking or simple environment.

SNA — SNA is a proprietary IBM architecture and set of products for implementing network computing within an enterprise. SNA is a mature protocol and is primarily oriented toward mainframe and mainframe peripheral communications. It is now part of IBM's Systems Application Architecture (SAA) and Open Blueprint. SNA has been withering for a long time but is not yet dead.

Software-Defined Network (SDN) — An AT&T product. According to AT&T, "Software Defined Network — International (SDNI) — is a high-quality virtual networking service providing worldwide connectivity for voice, fax, and data through point-to-point, two-way voice, and voice-band data communications." SDN is not, from the customer perspective, a typical packet-based VPN; calling SDN "virtual" may be true from a technical perspective, but it has caused confusion in the marketplace. If a customer makes long-distance calls using AT&T's SDN, the call still goes out over the PSTN and a per-minute charge still applies.

Switched Digital Video (SDV) — SDV system transports digital IP video over an IP network.

Switched Virtual Circuit (SVC) — SVC is a temporary virtual circuit. It exists only during the time that data is being transmitted. This is in contrast to the permanent virtual circuit that remains always up.

Symmetric DSL (SDSL) — SDSL uses a single twisted-pair line and supports speeds of 1.544 Mbps (United States and Canada) — or 2.048 Mbps (Europe) — in each direction on a duplex line. It is symmetric because the data rate is the same from either direction.

Synchronous Optical Network (SONET) — SONET is a Layer 1 (physical) protocol. It defines a hierarchy of interface rates that allow data streams at different rates to be multiplexed. SONET includes speeds of 51.8 Mbps up to 2.48 Gbps.

T1 Carriers — T1s were the first digital lines. They use two wire pairs; a total of 24 channels are available. Each channel can use either 64 kbps or a combination of 8 kbps for signaling and 56 kbps for voice/data.

Tandem Switching — An architecture in which one trunk is connected to another ("in tandem"). Typically, a tandem switch is between an originating switch and a final destination switch. The tandem switch merely passes the call along.

Telecommunications Industry Association (TIA) — The Telecommunications Industry Association is a trade association in the communications and information technology industry that facilitates business development opportunities and a competitive market environment. It has more than 1100 constituent companies that manufacture or supply the products and services used in global communications.

Terminal Equipment — Any device that terminates a communications channel and adapts that channel for use by a user, the user being either a person or a machine. Telephone sets, switchboards, datasets, answering sets, etc. are examples of terminal equipment.

Tie-Lines — Dedicated connections between sites within an organization.

Time Division Multiplexing (TDM) — TDM is a method of transmitting data or voice traffic. It allows users to take turns in a round-robin fashion, where each one periodically gets the entire bandwidth for a little burst of time (compare to FDM). It can be handled entirely by digital electronics but data must be digital; each voice channel requires a data rate of 64 kbps. Increasingly, the term TDM is used to designate older circuit switched technology in contrast to newer, IP-based systems.

Toll-Free Numbers — There are currently four toll-free prefixes in use — 800, 888, 877, and 866 — with more than 24 million toll-free numbers assigned. Other numbers are on the drawing board.

Total Cost of Ownership (TCO) — Lifetime costs of any system, device, or software. Includes initial purchase, installation costs, routine maintenance and updates, and eventually disposal costs (if applicable).

Transmission Control Protocol/Internet Protocol (TCP/IP) — TCP/IP is used to connect computers on the Internet. TCP/IP uses several protocols, including, of course, TCP and IP. TCP/IP is the *de facto* standard for transmitting data over networks. Many operating systems, such as Windows XP, UNIX, and Linux support TCP/IP.

Trunk — A transmission path usually used as a common artery between switching units, switching centers, toll centers, test centers, PBXs, and concentrators.

Uniform System of Accounts (USOA) — The classification by accounts of telephone plant, income, operating revenues, operating taxes, etc, as property costs, revenues, expenses, etc. This classification is prescribed by the FCC for all established telephone common carriers.

United States Telecommunications Association (USTA) — A broad-based association for local exchange carriers.

Universal Resource Locator (URL) — A URL is the address of a file (resource) accessible on the Internet. Servers on the Internet resolve the "English" name to an IP address.

Usage Sensitive Pricing (USP) — A generic term that includes measurement of local exchange service and directory assistance charging. It also includes any pricing that varies directly by usage (e.g., cents per minute).

Value-Added Network (VAN) — A network using the communication services of other commercial carriers, using hardware and software that permit enhanced telecommunication services to be offered.

Very High Bit Rate DSL (VDSL) — Transmits data in the 13- to 55-Mbps range over short distances, somewhere between 1000 and 4400 feet. It goes over copper wire and the speed is inversely proportional to the distance.

Very Small Aperture Terminal (VSAT) — VSAT technology uses a relatively small dish, aimed at a satellite 22,000 miles above the Earth, to communicate with a terrestrial "hub" that is in turn linked by "backhaul" circuits to an office or series of locations.

Virtual Private Network (VPN) — A virtual private network (VPN) is entirely private but is not part of one organization's infrastructure. A VPN service provider offers a publicly available network that provides full IT support. By encapsulating the data with encryption, third parties cannot see the information being "tunneled" over the Internet between the VPN and the organization's firewalls.

Voice-over-ATM (VoATM) — ATM has always had the capacity to carry delay-sensitive traffic and guarantee quality of service. Thus, Voice-over-ATM is akin to saying that "water is wet" — from its inception, ATM was designed to be a universal carrier of voice, data, fax, and video.

Voice-over-DSL (VoDSL) — According to Commweb.com, "Sending voice over a DSL line. Using compression, a large number of voice channels can be placed on DSL channels, which makes the technology very attractive. For example, up to 150 voice channels can be transmitted over a 1.5-Mbps DSL line. DSL signals at the customer side are delivered into an integrated access device (IAD), which forwards them over twisted pair to the carrier. The signals go to the carrier's DSLAM and then to an access switch that forwards voice to a voice gateway and then the PSTN and data to the appropriate data network."

Voice-over-Frame Relay (VoFR) — With an adequate CIR, limitations of frame sizes and other modifications of Frame Relay parameters, reasonably good voice traffic can be sent over Frame Relay. From a Frame Relay, Layer 2 perspective, voice is just another set of packets to be put into the payload portion of the frame.

Voice-over-IP (VoIP) — VoIP uses, obviously, IP to carry voice traffic. Because IP is a Layer 3 protocol, any number of Layer 2 protocols can support it. For example, VoIP can be supported by ATM, DSL, traditional T1s, or Frame Relay. In VoIP, voice information is digitized into discrete packets and sent as "data." VoIP uses network resources more efficiently than traditional circuit switched protocols because it does not take up bandwidth when it is not needed. One advantage of VoIP and Internet telephony is that it avoids the tolls charged by ordinary telephone service.

Voice Response Unit (VRU) — Also called "IVR," for interactive voice response. These are the ubiquitous "press 1 for account balance information, 2 for your most recent transactions" systems that provide automated information over the telephone network.

Wavelength Division Multiplexing (WDM) — Wavelength division multiplexing (sometimes called dense WDM) combines data from various sources and funnels them onto an optical fiber. Each signal has its own separate color (wavelength). Up to 80 channels can be multiplexed on to a single optical fiber. Up to 2.5 gigabits per second per channel can be transmitted and up to 200 Gbps for the fiber. Within the channel, the signal is TDM.

Wide Area Network (WAN) — A WAN is internal, similar to a LAN, but includes a much larger geographic scope — another city, state, or country. The designation of WAN has nothing to do with the layer 2 technology. Telephone lines, satellite, Frame Relay, wireless third-generation links, or any other long-distance architecture can support it.

Wide Area Telecommunication Service (WATS) — A service for customers having substantial volumes of interstate long-distance calls over a wide area. Rates are based on total usage as opposed to a call-by-call basis. 800 service enables customers to receive calls from selected service areas, without charge to the calling party.

Wireless Application Protocol (WAP) — WAP is an umbrella specification for a set of communication protocols to standardize the way that wireless devices, such as cellular telephones and radio transceivers, can be used for Internet access, including e-mail, the World Wide Web, newsgroups, and other forms of communication. WAP enables Internet content developers to consistently deliver information formatted for the smaller screens characteristic of most wireless devices.

Wireless Code Division Multiple Access (WCDMA) — WCDMA is officially known as IMT-2000 direct spread. Internally, it works much like wire line Ethernet, using collision detect algorithms. It is a third-generation (3G) mobile wireless technology offering higher data speeds to mobile and portable wireless devices than commonly offered in existing technologies.

Wireless Local Loop (WLL) — A local loop is the wired connection from a telephone company's central office in a locality to its customers' telephones at homes and businesses. In the past, this connection was usually accomplished with a pair of copper wires called twisted pair. With the wireless loop, appropriate wireless transmitters, receivers, and cards are used to accomplish the same purpose.

Working Telephone Number (WTN) — This represents an actual telephone number that can be dialed. Large organizations will have a "bill to" number that may be associated with trunks but cannot be dialed directly.

Appendix H

Audit Program for Telecom Cost Management Field Work

The following material assumes prior reading of Chapter 3. Although there is some repetition from Chapter 3, *this audit program is structured for those doing the work.* The instructions are from the perspective of a third-party billing audit but the substitution of "your organization" for "the client" is all that is required for use internally.

Telecommunications is changing rapidly and telecommunications billing naturally follows suit. Terms, USOCs, and offerings change daily. However, even as organizations converge their data, voice, wireless, and other platforms, the billing issues remain relatively constant. We recommend using this audit program as a *starting point.* You should identify any new technologies or special circumstances and develop a new program tailored to your organization.

1. Detailed Steps: Flowchart Description

The following major steps are involved in the Planning phase.

1.1 *Finalize project documents and scope.* Finalize project scope, identify selected sites and vendor contacts, and obtain records.
1.2 *Obtain stakeholder buy-in.* Communicate project to all impacted personnel.
1.3 *Request documentation.* Allow four (4) to six (6) weeks for documentation to arrive.

1.4 *Verify receipt of all requested documentation and check to make sure that it is complete.* Upon receipt of documentation, review to ensure that material includes appropriate detail.

1.5 *Evaluate need for site visit.* Review necessity of site visits to inventory switch detail.

1.6 *Conduct kick-off.* Bring project team together to officially begin project.

1.1 Finalize Project Documents

Before you begin

Understand any team member or management expectations before finalizing project content.

Finalize project scope

Finalization of the project scope contains the following elements:

1. Finalize project content/scope:
   a. Enterprisewide or limited focus
   b. Voice and/or data
   c. Local service, IXC service, and/or equipment
   d. Strict recovery or value-added
   e. Network and/or equipment
2. Set project timeline and staffing requirements.
3. Determine project facility requirements (analog lines, phones, desks).
4. Clarify requirements and penalties, if any, for not meeting those requirements (i.e., materials must be reviewed within five days or project will be delayed two weeks).
5. Clarify expectations.
6. Create sample final deliverables.

Identify sites/contacts

Finalization of the project scope contains the following elements:

1. Solidify selection of sites and identify site contacts with client management.
2. Meet regional telecom leaders.
   a. Document bill processing procedures.
   b. Identify key staff.
   c. Identify carrier contacts.
   d. Determine carrier issues.
   e. Obtain copies of contracts.
   f. Obtain any maintained detailed billing data:
      (1) Lines
      (2) Trunks

       (3) Circuits
       (4) Equipment:
           (a) Switches
           (b) Key equipment
           (c) Routers
           (d) Channel banks
       (5) Obtain circuit topology

3. Learn A/P process, determine whether functions are centralized or decentralized, and obtain A/P telecom vendor maintenance names and numbers:
   a. Accounts Payable and General Ledger information
      (1) Vendor names and contact information
      (2) Names and phones numbers of contacts who impact billing process
      (3) Annual or monthly spending
      (4) Types of invoices or items coded to telephone expense codes
      (5) Chargeback policies
      (6) Internally held billing information
         (a) Where are the records held? Centrally or at each site?
         (b) Are they accessible?
         (c) Is workspace near the records available?
      (7) Improper identification of vendors can delay project
4. Learn provisioning process, and determine whether provisioning is centralized or decentralized.
   a. Obtain any available documentation
   b. Obtain contact names and numbers of vendor provisioning staff

1.2 Obtain Stakeholder Buy-In

Before you begin

Have ready the final project details write-up.

Obtaining buy-in

Sample steps to obtain buy-in (every organization is different in how initiatives are conducted):

1. Update any revised project details.
2. Review all of project details emphasizing that project cannot start until all billing documentation is received (typically four to six weeks after request).
3. Agree upon project deliverables before project begins.
4. Communicate, via meeting or notice, project purpose, plan, expected results, level of assistance required, etc.

1.3 Request Documentation

Before you begin

Obtain a complete list of pertinent telecom vendors with contact information. Relying on a single reference source will usually result in incomplete records.

Exhibit 1.   Types of Information Used in a Telecom Billing Audit

Item	Source	Use
Directories	In-house	Office addresses, phone numbers, and telecom or billing contacts
General ledger	Financial management	Vendor account information
Inventories Business lines Circuits	Telecom group	Lines or circuits, for each location, possibly containing addresses and purpose
Trunk reports	Telecom group	PBX printout of trunk groups
Customer service records (CSRs)	LEC	Detailed itemization of monthly recurring charges and listing of lines and billing/service addresses
Invoices — paper	Vendor	Monthly invoices for telephone service purchased
Billing CD-ROM (if available)	Vendor	Monthly invoices, regional or firmwide. Files on CD include call detail, circuit information, recurring charges, and database with firm's numbers.
Contract information	Vendor or in-house	Contractual recurring charges by vendor
Switch room	Telecom group	Can verify which lines/circuits come into facility

Multiple sources of material must be cross-checked to ensure completeness of material.

Types of data

The types of information listed in Exhibit 1 will be used in a Telecom Billing Audit.

Requesting documentation

Follow these steps to request material and to document receipt of material.

1. One team member should be responsible for requesting, tracking, and ensuring completion of all requests. This person will be the Documentation Lead.
2. All bills should be directed to the location of the Documentation Lead.
3. Obtain PBX reports for applicable locations.
4. The following list of vendor materials can be obtained from the Accounts Payable department. This list should be modified to reflect the scope of project (e.g., if project does not include PBX, then equipment contracts and maintenance agreements do not need to be included):

    a. Vendor invoices:
- (1) Customer service records (CSRs) to determine lines billed (Client may not receive this documentation.)
- (2) LEC or CLEC all services
- (3) IXC all services
- (4) Telephone equipment vendors for purchase, lease, maintenance, moves, adds, and changes
- (5) Calling card
- (6) Two-way radio
- (7) Pagers
- (8) Cellular
- (9) Data

    b. Vendor contracts:
- (1) Contracts with LEC or CLEC for local service
- (2) Contracts with long-distance providers
- (3) Lease agreements for equipment
- (4) Maintenance agreements for equipment (This is a key area for potential savings opportunities.)

Documentation: Directories

Purpose

An internal directory provides information on telephone numbers used and their purpose. Look for a directory containing information on the DID range, if possible. Information on site addresses for a multi-site engagement can be extracted from a corporate directory. Also, toll-free numbers, and their function, can often be found in the company directory (see the example in Exhibit 2).

Use

There may be more than one internal directory. Each directory is created for a specific location or region.

    The accuracy of a directory should always be questioned. The age of the directory should be determined as well as how often it is updated. The availability of information to the individual who prepares the directory should also be assessed.

Documentation: General Ledger

Purpose

A general ledger can provide information on account numbers for various vendors. The general ledger provides spend detail and can be used to determine additional telecom vendors not identified (see the example in Exhibit 3).

Exhibit 2. Sample Company Directory

Full Name	Street Address	City	State	Zip	Office Phone Number	Office FAX Phone Number
R. Accomando	8916 Club Creek	Dallas	TX	75238		
Gay Addington	931 W. Colorado	Dallas	TX	75208		
Jackie Blair	One Bell Plaza					
Karen Clapp	108 Bush	La Jolla	CA		619-239-0984	
Jan Cummings	311 Akard	Dallas	TX		972-889-5805	972-889-5806
Margaret Dionne	2035 Lavaca Trail	Carrollton	TX	75010	972-774-8603	
D. Greiner	8500 Texas Ave.	San Antonio	TX		800-872-2872	
L. Jacobs	2300 W. Broadway Ave.	Ft. Worth	TX	76102	817-737-4722	
Laura Spurgin	2085 Midway	Dallas	TX			

Exhibit 3. Sample General Ledger

Vendor Name	Invoice Num	Invoice Date	Amount	Description
INTERCALL	246460	May-99	1444.42	AC#018546/CONF CALLS 3/98
MCI TELECOMMUNICATIONS	71617610	May-99	1735.78	LONG-DISTANCES TELEPHONE CHARGES
NETWORKMCI CONFERENCING	080198 $591.67	May-99	591.67	AC#01-00026259026-00970
PACIFIC BELL	120197 $101.11	May-99	101.11	LA-ACCT #250 128 9947 245 159 N 6
PACIFIC BELL	120197 $149.88	May-99	149.88	LA-ACCT #250 128 2666 734 159 N 3
PACIFIC BELL	120197 $198.30	May-99	198.30	LA-ACCT #250 128 9946 841 159 N 3
PACIFIC BELL	112097 $285.24	May-99	285.24	LA-ACCT #650 473 8975 620 159 N 9

Exhibit 4.   Circuit Inventory

Circuit No.	From	To	Purpose	Function	Telephone Number
11HCQA988839	HOPE	CARRIER	IXC	VOICE	2132360544
11TLNA7162646	GRAND	NEWPORT BEACH	TIE	VOICE	
36KADA715495	HOPE	CENTURY CITY		56K	
11HCQA967375	HOPE	LEC	LOCAL	VOICE	2136124427

Exhibit 5.   Business Line Inventory

Telephone Number	Location	Function
415-444-1299	9817 Main, Grand Banks, OK	Modem
415-344-9999	9817 Main, Grand Banks, OK	Private line, Mr. Zed
415-344-1212	9817 Main, Grand Banks, OK	Time card machine
415-345-6767	9815 Main, Grand Banks, OK	Security guard

Use

If the telephony spend is not specifically categorized, the information is more difficult to extrapolate. The most recent, complete vendor invoice needs to be supplied.

Items to note:

- The vendor, the item, and spend
- Vendors found here that were not identified
- Account numbers by vendor should be noted only if the engagement is not firmwide and the general ledger breaks down the charges by location
- Vendor bills for nontraditional items should be recorded; specifically, AT&T Small Business bills need to be identified separately from AT&T products, which are contractually obtained.

Documentation: Circuit or Business Line Inventories

Purpose

Inventories include any documentation that reflects lines and circuits. The documentation may include identifier, purpose, and physical location (see Exhibits 4 and 5).

Use

The accuracy of any maintained listing should always be questioned. The age of the listing should be determined as well as how often it is updated. The availability of information to the individual who prepares the listing should also be assessed.

Exhibit 6. Sample List Trunk-Group

Group No.	No. TAC	Out Queue Group Type	Group Name	Mem	TN	COR	CDR	Meas	Disp	Len
1	601	isdn-pri	dcs pri dr3:dr6	8	1	1	y	none	n	0
2	602	cama	emergency 911	2	1	1	y	none	n	0
3	603	tie	dcs ds1 dr3:dr6	4	1	1	y	none	n	0
4	604	tie	dcs data	1	1	1	y	none	n	0
5	88	tie	dr2/ds1-data	1	1	1	y	none	n	0
6	86	tie	dr2/ds1-tie	4	1	1	y	none	n	0
7	87	tie	dr22/dr3-ds1	4	1	1	y	none	n	0
8	608	isdn-pri	dr22/pri	4	1	1	y	none	n	0
9	609	tie	DCS/dr22/data	1	1	1	y	none	n	0
10	610	isdn-pri	PRI/dr22	4	1	1	y	none	n	0
11	611	isdn-pri	PRI/dr06	9	1	1	y	none	n	0
12	612	isdn-pri	DTGS-DCS	4	1	1	y	none	n	0
13	613	tie	DTGS-DCS	4	1	1	y	none	n	0
14	614	tie	DCS dr3:dr6 alg	1	1	1	y	none	n	0

Documentation: PBX Trunk Reports

Trunk reports from the switch, or PBX, are used to analyze trunk activity and trunk status. This information discloses characteristics of the trunk groups programmed and connected to the switch that may affect billing (see Exhibits 6 through 12).

Procedure

Follow these steps to request trunk reports for Avaya switches. Trunk reports for other manufacturers will be similar.

Reports

- List of Trunk Groups, with purpose and quantity
- Measurements of Use of Trunk Groups, or Members of Trunk Groups

Supplemental Information

- Trunk Listing
  - List Trunk-Group*
  - Display Trunk-Group <trunk group number>

---

* These commands are specific to the Avaya Definity PBX. Other PBXs, such as Nortel's Meridian series, will have equivalent commands. The resident switch technician or maintenance vendor can typically run these reports for you.

TRUNK GROUP   Page Y of X

Administered Members (min/max): xxx/yyy

Total Administered Members: xxx

Group Member Assignments

	Port	Code	Sfx	Name	Night	Mode	Type	Ans Delay
1:	_____	_____	____	_____	_____	_____	_____	
2:	_____	_____	____	_____	_____	_____	_____	
3:	_____	_____	____	_____	_____	_____	_____	
4:	_____	_____	____	_____	_____	_____	_____	
5:	_____	_____	____	_____	_____	_____	_____	
6:	_____	_____	____	_____	_____	_____	_____	
7:	_____	_____	____	_____	_____	_____	_____	
8:	_____	_____	____	_____	_____	_____	_____	
9:	_____	_____	____	_____	_____	_____	_____	
10:	_____	_____	____	_____	_____	_____	_____	
11:	_____	_____	____	_____	_____	_____	_____	
12:	_____	_____	____	_____	_____	_____	_____	
13:	_____	_____	____	_____	_____	_____	_____	
14:	_____	_____	____	_____	_____	_____	_____	
15:	_____	_____	____	_____	_____	_____	_____	

Exhibit 7.   Sample Display Trunk-Group

■ Report Types
 – List Measurements Trunk-Group Hourly <trunk group number>
 – List Trunk-Group Summary Today
 – List Measurements Trunk-Group Summary Last
 – List Measurements Trunk-Group Summary Yesterday
 – List Performance Trunk-Group Today
 – List Performance Trunk-Group Yesterday

Switch Name:_____   Date: 4:10 pm TUE SEP 7, 2001

TRUNK GROUP HOURLY REPORT

Grp No: 87   Grp Size: 48   Grp Type: did   Grp Dir: inc   Que Size: 0

Meas Hour	Total Usage	Maint Usage	Total Seize	Inc. Seize	Tandem Seize	Grp Ovfl	Call Qued	Que Ovf	Que Abd	Out Srv	% ATB	%Out Blk
300	6	0	2	2	0	0	0	0	0	0	0	*
200	2	0	5	5	0	0	0	0	0	0	0	*
100	4	3	10	10	0	0	0	0	0	0	0	*
0	2	0	5	5	0	0	0	0	0	0	0	*
2300	5	0	8	8	0	0	0	0	0	0	0	*
2200	7	0	10	10	0	0	0	0	0	0	0	*
2100	16	0	29	29	0	0	0	0	0	0	0	*
2000	16	0	22	22	0	0	0	0	0	0	0	*
1900	31	0	16	16	0	0	0	0	0	0	0	*
1800	26	0	27	27	0	0	0	0	0	0	0	*
1700	36	0	27	27	0	0	0	0	0	0	0	*
1600	23	0	21	21	0	0	0	0	0	0	0	*

Exhibit 8.   Sample List Measurements Trunk-Group Hourly (for Trunk-Group Number) 87

Documentation: LEC Information

Before you begin

Obtain the following lists:

- Working telephone numbers and associated service addresses
- Circuit numbers and service addresses
- Equipment type and service addresses
- LEC contacts and phone numbers
- PBX reports

Enter information

Follow these steps to enter the information.

Switch Name:_____ Date: 4:11 pm TUE SEP 7, 2001

TRUNK GROUP SUMMARY REPORT

Peak Hour For All Trunk Groups: 1400

Grp No.	Grp Siz	Grp Type	Grp Dir	Meas Hour	Total Usage	Total Seize	Inc. Seize	Grp Ovfl	Que Siz	Call Qued	Que Ovf	Que Abd	Out Srv	% ATB	%Out Blk
26	46	isdn	two	1500	0	0	0	632	0	0	0	0	46	100	100
33	46	isdn	two	1500	209	163	86	0	0	0	0	0	0	0	0
82	10	co	inc	1500	51	32	32	0	0	0	0	0	0	0	*
83	48	tie	two	1500	1002	568	170	0	0	0	0	0	0	0	0
84	23	co	two	1500	469	227	0	0	0	0	0	0	0	0	0
85	12	co	two	1500	6	0	0	0	0	0	0	0	0	0	0
87	48	did	inc	1500	632	397	397	0	0	0	0	0	0	0	*

Exhibit 9.   Sample List Measurements Trunk-Group Summary Last

Switch Name:_____ Date: 4:10 pm TUE SEP 7, 2001

TRUNK GROUP SUMMARY REPORT

Peak Hour For All Trunk Groups: 1400

Grp No.	Grp Siz	Grp Type	Grp Dir	Meas Hour	Total Usage	Total Seize	Inc. Seize	Grp Ovfl	Que Siz	Call Qued	Que Ovf	Que Abd	Out Srv	% ATB	%Out Blk
26	46	isdn	two	2300	0	0	0	29	0	0	0	0	46	100	100
33	46	isdn	two	900	9	5	2	0	0	0	0	0	0	0	0
82	10	co	inc	1500	2	3	3	0	0	0	0	0	0	0	*
83	48	tie	two	1400	75	55	17	0	0	0	0	0	0	0	0
84	23	co	two	1400	54	10	0	0	0	0	0	0	0	0	0
85	12	co	two	2300	0	0	0	0	0	0	0	0	0	0	0
87	48	did	inc	1100	43	25	25	0	0	0	0	0	0	0	*

Exhibit 10.   Sample List Measurements Trunk-Group Summary Yesterday-Peak

Switch Name:_____ Date: 4:11 pm TUE SEP 7, 2001

HIGHEST HOURLY TRUNK GROUP BLOCKING PERFORMANCE

Grp No.	Grp Type	Grp Dir	Grp Size	% Outgoing Blocking or % ATB 1 2 3 4 5 6 7 8 9 10 20 30 40 50	%Out Blkg	%Time ATB	Meas Hour	Total Calls
26	isdn	two	46	*****************************	100	100	1500	0
33	isdn	two	46		0	0	1500	163
82	co	inc	10		*	0	1500	32
83	tie	two	48		0	0	1500	568
84	co	two	23	*****	3	2	1400	276
85	co	two	12		0	0	1500	0
87	did	inc	48		*	0	1500	397

Exhibit 11. **Sample List Performance Trunk-Group Today**

Switch Name:_____ Date: 4:11 pm TUE SEP 7, 2001

HIGHEST HOURLY TRUNK GROUP BLOCKING PERFORMANCE

Grp No.	Grp Type	Grp Dir	Grp Size	% Outgoing Blocking or % ATB 1 2 3 4 5 6 7 8 9 10 20 30 40 50	%Out Blkg	%Time ATB	Meas Hour	Total Calls
26	isdn	two	46	*******************************	100	100	2300	0
33	isdn	two	46		0	0	2300	0
82	co	inc	10		*	0	2300	2
83	tie	two	48		0	0	2300	35
84	co	two	23		0	0	2300	0
85	co	two	12		0	0	2300	0
87	did	inc	48		*	0	2300	8

Exhibit 12. **Sample List Performance Trunk-Group Yesterday**

1. Sort the material for the site.
2. Enter the location name, address, and billing telephone number (BTN) into the database.
3. Enter the local exchange carrier (LEC) name.
4. Enter the equipment type.
5. Enter BTN if known.
6. Enter the working telephone numbers (WTNs) for the site.
7. Or enter the circuit numbers for the site.

Documentation: LEC Customer Service Records and Invoices

Before you begin

Create a vendor request letter to obtain: local exchange carrier (LEC) Customer Service Records (CSRs) detail monthly recurring charges. Business line paper or CD-ROM invoices may include identifier, purpose, and physical location. Most firms will have paper bills. LECs do offer invoices and CSR detail on CD-ROM, but few use this service to the fullest extent. If CD-ROM exists in-house, verify that all services are represented. If not, obtain paper. Try to obtain all invoices and CSRs for the same time period.

Request CSRs and bills

Follow these steps to obtain CSRs and invoices for each site:

1. Identify the major account representative or the Vendor Liaison Group for the LEC.
2. Contact the LEC representative and obtain her fax number.
3. Fax the LOA to the LEC.
4. Fax or e-mail the list of WTNs and known BTNs with the associated service addresses and the firm's name or versions of names to the LEC and request:
    a. CSRs
    b. A recent bill or bills (depending on the scope)
    c. *Note:* Accept no substitutes for the actual CSRs.
    d. Request that all documentation of services at specific sites be provided:
    e. The BTN and WTN are both provided to ensure that potentially unknown numbers are subsequently identified and the documentation is received.
5. Enter into the database the date the fax or e-mail was sent in.
6. Follow up with the LEC until the information is received.
7. Enter the date of CSR and/or bill receipt into the database.
8. File the CSRs and bills by BTN.

Note: Not all LECs use the term "CSR." Southwest Bell, for example, calls them "SO" reports.

Documentation: Tariff and Contract Documents

Purpose

Gather documentation that supports the proposed rates for all call types. The tariffs should include state tariffs, interstate tariffs, and international tariffs.

Type	Origination	Termination	Rate	Rate Schedule
Interstate	Dedicated	Switched	$0.00	
	Switched	Switched		
	Dedicated	Dedicated		
	Switched	Dedicated		
	Toll Free Number	Dedicated		
	Toll Free Number	Switched		
Intrastate State Specific	Dedicated	Switched		
	Switched	Switched		
	Dedicated	Dedicated		
	Switched	Dedicated		
	Toll Free Number	Dedicated		
	Toll Free Number	Switched		

Exhibit 13. Suggested Rate Type Information

Include state tariffs for the states where the client has locations. The international tariffs naturally reflect charges for calls to all countries that the client calls.

Look for separate switched and dedicated access rates. In addition, toll-free rates and advanced features will have separate charges.

Special contracts include services such as:

- Revenue Commitment Levels
- Discount structures
- Length of contract
- Special rate indicators

Rates can be found in tariffs and are typically discounted based on a contract, or specific rate elements (see Exhibit 13) can be incorporated in a contract.

Tariff

Each carrier has multiple tariff offerings. All tariffs are filed with the FCC and contain pricing for interstate and international service offerings. Tariffs are also filed with state Public Utility Commissions for intrastate products and services.

Exhibit 14 represents a sample AT&T Intrastate Tariff.

AT&T COMMUNICATIONS OF
STATE SPECIFIC , INC.

State P.U.C.-No. 10
Section 2
First Revised Sheet 1

SAMPLE CALLING PLAN SERVICES

2.1.   AT&T PRO STATE SPECIFIC

   2.1.1.   General

   AT&T PRO State Specific is provided to customers who place customer-dialed, AT&T station calls within the State of State Specific . For a fixed monthly price, customers may make any number of such calls at a discount from the normal price charged for those calls as specified in 2.1.3.C. following.

   2.1.2.   Regulations

   A. This plan is provided only where facilities and billing capability permit. Any call billed under this plan may not receive further discounts under any other plan.

   B. The discount offered under this plan applies to the total of all eligible usage during a billing period, not on individual messages. Usage of applicable calls will be totaled and then discounted from the price schedule as specified in 2.1.3.C. following.

   C. All lines and trunks billed to the same billing account, except for services in foreign central offices, are included in the plan. This also applies when consolidated billing is provided by the billing agent.

   D. AT&T PRO State Specific does not apply to calls other than customer-dialed, station calls including, but not limited to:
   1. Conference Service calls
   2. Directory Assistance Service
   3. Customer Dialed Calling Card Station, Operator Station, Person-to-Person, and Real Time Rated

   E. Customers who retain service for less than one month will be billed the full per month charge for the plan s applicable usage charges. Beyond the first month when customers choose to no longer retain the service, per month charges will be prorated for a partial month as appropriate.

OPTIONAL CALLING PLAN SERVICES

   2.1.3.   Prices

   A. Monthly

	Price	USOC
AT&T PRO State Specific per account	$ 5.00	TS1LV
Usage Discount	10%	

   B. Nonrecurring

Service Order charge, per account#	$10.00

   AT&T Communications of State Specific may waive this charge. The nonrecurring service order charge will not apply when an AT&T PRO State Specific customer

---

Exhibit 14.   Sample Calling Plan Services

moves from its present location and, within 30 days, subscribes to AT&T PRO State Specific at its new location.

C. Price Schedule

The following table contains the first minute and additional minute prices for the Day, Evening and Night/Weekend periods for AT&T PRO State Specific customer-dialed, station calls. These prices are based on chargeable time (duration) of the message, as specified in 2.1.5. following. The applicable airline distance between rate centers is determined as provided in AT&T Tariff F.C.C. No. 10.

Mileage	Day		Evening		Night/Weekend	
	Initial Minute	Each Add 1 Minute	Initial Minute	Each Add 1 Minute	Initial Minute	Each Add 1 Minute
1 10	$.2475	$.1350	$.2422	$.1211	$.2040	$.1020
11 22	.2475	.1350	.2422	.1211	.2040	.1020
23 55	.2700	.1507	.2636	.1496	.2220	.1260
56 124	.2970	.1890	.2850	.1781	.2400	.1500
125 29 2	.3240	.1980	.3064	.1924	.2580	.1620
293 35 4	.3420	.2250	.3206	.2137	.2700	.1800

D. Initial Period and Additional Minutes

Prices are quoted in terms of initial period and additional minutes.

1. Initial period prices given in the price schedule in 2.1.3.C. above are for connections of one minute or any fraction thereof.
2. All additional minute prices given in the price schedule in 2.1.3.C. above are for each additional minute or any fraction thereof that the connection continues beyond the initial period.

E. Directory Assistance

AT&T PRO calls for Directory Assistance Service placed by customers of this illustrative company are subject to the regulations and prices for such service as provided in Tariff State P.U.C.-No. 8, Section 5. PRO usage charges do not apply to calls made to Directory Assistance.

2.1.4. Time-of-Day and Day-of-Week

The price schedule shown in 2.1.3.C. applies as follows:

- *Day* From 8:00 A.M. TO 5:00 P.M., Monday through Friday.
- *Evening* From 5:00 P.M. to 11:00 P.M., Monday through Friday and Sunday.
- *Night* From 11:00 P.M. to 8:00 A.M., every day; from 8:00 A.M. to 11:00 P.M. on Saturday and from 8:00 A.M. to 5:00 P.M. on Sunday.

2.1.5. Chargeable Time

A. The time when connection is established, as provided in B. following, determined in accordance with the time, standard or daylight saving, observed at the location of the rate center of the calling station, determines what price schedule applies. In cases where a message begins in one price period and ends in another, the charge for the portion of the message within each price period shall be the charge for the whole minutes in effect for that period.

Exhibit 14.   Sample Calling Plan Services (Continued)

    B. Chargeable time for customer-dialed, station calls begins when connection is established between the calling station and the called station, miscellaneous common carrier mobile radio system, PBX or Centrex system and ends when the calling station hangs up thereby releasing the network connection. If the called station hangs up but the calling station does not, chargeable time ends when the network connection is released either by automatic timing equipment in the telecommunications network or by the Company operator.

    C. Chargeable time does not include time lost because of faults or defects in the service.

Exhibit 14.   Sample Calling Plan Services (Continued)

Documentation: Toll-Free Numbers

Typically, there is no complete, up-to-date list of all toll-free numbers and the telephone number to which they are routed or terminated. Several information sources will be used to build a site-by-site list of toll-free numbers.

Source

The following is a list of four sources for toll-free numbers.

- Accounts payable
    - Review general ledger entries for telecom vendors.
    - Look for account numbers that are toll-free numbers.
- Telephone directory
    - Review the telephone directory for toll-free numbers.
- Telecom group
    - Request a list of toll-free numbers per site.
- IXC customer representative
    - Ask the account team for a list of toll-free numbers and recent detailed bill.

1.4 Verify Receipt of Requested Documentation

Without data from all four types of information — the existing documentation, LEC bill, LEC CSR, and the IXC bill — savings or recovery opportunities may be significantly reduced.

    The Documentation Lead should ensure that the following is done.

1. One team member should be responsible for requesting, tracking, and ensuring completion of all requests.
2. All requests should be documented on the billing request matrix, which tracks date requested, date expected, and date received.
3. Upon receipt of each document, ensure that the bill is a detail bill rather than a summary bill. If it is a summary bill, determine if usage is provided

Exhibit 15.   Need for Additional Material

Issue	Source	Action
Terminating number is switched, but not in Customer Service Record Worksheets	Toll-Free Service	Request CSRs and LEC bills for service
PICCed lines are found in IXC documentation	PICCed Lines Worksheet	Request CSRs and LEC bills for service

in a separate format or media such as CD-ROM. If not, then contact vendor and request the detail information again.

4. After document has been verified, file with like bills by Billing Telephone Number.
5. Notify project manager if problems are encountered with getting the right type of bill.

Need for Additional Data

Once documentation has been received and reviewed, a need for additional material can be identified (see Exhibit 15). The process to obtain additional data will be ongoing and will follow the same flow as the initial steps for material gathering.

1.5 Evaluate Need for a Site Visit

A visit to the switch room may be required to inventory services and equipment pertinent to the engagement.

Criteria

The project manager should evaluate the following criteria to determine whether a visit to the switch room is required.

1. *Does the switch room support a service center or call center?* A call center would generate enough usage to justify a visit to the switch room.
2. *Has at least one site been visited during the project?* At least one switch room should be reviewed to verify that no major issues exist. Additionally, other problems may be detected, including:
   a. An unusual quantity of POTs lines that are not terminated in any equipment
   b. Lack of physical security for the PBX
3. *Are meetings scheduled at a site where a switch room is located?* If a trip is already planned to a site, you may want to also schedule a visit to the switch room.

4. *Is there a centralized or decentralized order provisioning process?* A centralized order provisioning process would allow you to rely on inventory documentation provided by the client.
5. *How many data lines terminate at the site?* A greater quantity of data lines would justify a visit to the switch room.

To conduct site visit

1. Locate equipment room(s).
2. Locate backboard or demarc point.
3. Document all lines and circuits identified on demarc — create documentation, note line or circuit number.
4. For services that are not properly labeled, identify line number:
   a. Obtain dial tone.
   b. Dial 1010–732–1770–988–9664.
   c. Listen for the message that contains the origination number.
5. Verify IXC.
   a. Obtain dial tone.
   b. Dial 1–700–555–4141.
   c. Listen to message.
   d. If carrier is AT&T, then dial 1–700–951–5555 to determine if number is on a specific AT&T plan.
6. Document IXC (PIC) — make note of the IXC.
7. Determine function or use. Note findings.

1.6 Conduct Project Kick-Off

Before you begin

Verify that all billing documents have been received in full.

Project kick-off

Follow these steps to officially kick off the project.

1. Distribute agenda prior to meeting.
2. Introduce key players.
3. Review project outline and requirements, describe potential pitfalls discovered thus far, and re-verify verbal agreement of project scope and deliverables.
4. Schedule meetings with team to present:
   a. Updates such as percent of completion
   b. Specific requests for assistance
   c. Scheduled date of deliverables

2. Entering the Data

The Enter Data phase involves entering the data received from the LEC and the IXC into the database for further analysis.

If the material is provided in soft form, via CD-ROM, the transfer of data may be mechanical. Although most vendors offer CD-ROM bills, many have not taken advantage of this service.

Data sources

- Customer service records: detailed itemization of the monthly recurring charges.
- Gathered information: worksheet.
- Toll-free invoices: toll-free worksheet, listing of lines or circuits possibly containing addresses and purpose.
- Invoices (paper): monthly invoices.
- CD-ROM: contains a copy of the monthly invoices, regional or firmwide. Files include call detail, circuit information, recurring charges, and database with firm numbers.
- Internal interviews: maintain notes in work folder.
- Meetings with vendors: maintain notes in work folder.

Flowchart

Exhibit 16 shows the major steps involved in this phase and the external entities that provide information into the process.

Flowchart description

The following major steps are involved in the Enter Data phase.

2.1 *Enter LEC CSR information.* Monthly recurring charges from the CSR are inputted into a database system.
2.2 *Enter LEC invoice information.* The LEC bill is reviewed; the total usage per number is entered into the database.
2.3 *Enter IXC billing information.* Telephone and circuit numbers are extracted by bill group. The numbers, bill group, PIC charge, and total usage are entered into the database.
2.4 *Enter firm information.* Firm data is entered into the database.
2.5 *Verify completeness of documentation.* Ensure that all missing data has been filled in for a complete view of a client's telecom billing.

2.1 Enter LEC CSR Information

Materials needed

The following materials are needed for this procedure:

- LEC CSRs

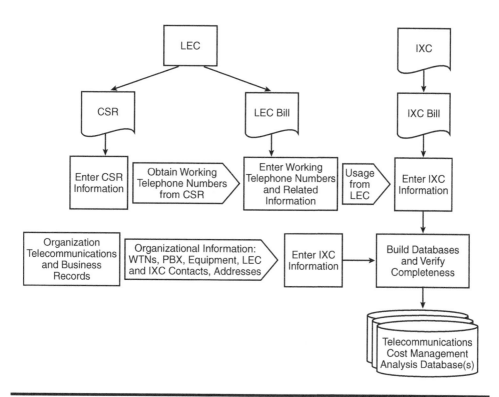

Exhibit 16.   Data Collection Process for Telecommunications Auditing

Enter CSR information

Follow these steps to enter working telephone number (WTN) and circuit information from the LEC CSR.

1. Enter the WTN into the database.
2. Enter an 'X' in the CSR Source field to indicate obtained data from the CSR.
3. Enter the total recurring charge associated with the number.
4. Enter the quantity of lines to calculate the recurring charge per unit.
5. Enter the PIC for the WTN.
6. Enter the service type in the Service Type field. Example: 1MB, 1FB, T1, or ISDN.
7. Is there a service address in the database for the number? (For circuits, also obtain the terminating service address.)
8. Add the new service address into the database.
9. Enter CSR in the Address Source field.
10. Select the appropriate service address from the Service Address field.
11. Initial the hardcopy CSR.

Data Entry: Customer Service Record

The customer service record (CSR) is a detailed itemization, by billing telephone number (BTN), of the monthly recurring charges from the LEC (local exchange carrier).

Exhibit 17.   CSR Data on Project Worksheets

Worksheet	Location on CSR	Description
Business line	Equipment	WTN appears in a separate field or in the body of the CSR Description
Trunk	Trunk	Individual or pilot trunk numbers will appear in a separate field or in the body of the CSR Description
Circuit	Equipment	Circuit Number contains alpha and numeric characters; appears in a separate field or in the body of the CSR Description
DID	Common equipment or system services	Range of numbers assigned to PBX stations for inbound delivery of calls

Worksheets

CSR data will populate one of four project worksheets (see Exhibit 17).

CSR Elements

- *Listings*: white page listings
- *Bill*: address where the monthly bill is mailed
- *Rmks*: remarks section
- *Equipment*: detail on individual lines or hunt groups and itemized charges, **or**
- *Services and features*: circuit detail information, **or**
- *Common equipment, or system services*: PBX services such as DID blocks **and**
- *Trunks*: detail on trunks and the itemized charges

CSR Codes

The following table describes CSR codes commonly viewed in a billing audit project. However, these will vary somewhat, depending on the carrier.

- *1FL.or 1FB*: single business line
- *1HB.*: one measured business line
- *ND8*: first block of 100 DID numbers
- *ND9*: subsequent block of 100 DID numbers
- *TFC*: trunk
- *TFB02*: one-way incoming
- *TFB02*: one-way outgoing
- *TTB*: touch-tone
- *9ZR*: FCC-approved customer line
- *RCFVF*: telebranch

- *T15CX*: T1 line
- *TFB*: combination trunk
- *TFB02*: DID (directly inward dialed)
- *TFB03*: combination trunk
- *TFC*: combination trunk
- *UFKB#*: ISDN

CSR Data Entry: Business Line Worksheet

Information from the *Equipment* section of the LEC CSR will be entered into this worksheet. Exhibit 18 shows a sample business line worksheet, and Exhibit 19 displays the source document.

Worksheet Fields

- **BTN** (main billing number): account number or telephone number
- **WTN** (working telephone number): individual line (the BTN is one WTN)
- **Service Address** (location): termination of service (except remote call forwarding service)
- **LEC** (local exchange carrier or local telephone company): incumbent local operating company
- **Monthly Recurring Charge:** total monthly billing of all WTN elements
- **LPIC** *or* **ZPIC** (intraLATA toll provider): in areas where the intraLATA toll provider can be other than the local operating company
- **PIC** (interLATA-intrastate, interstate, international toll provider): four-digit code indicating preferred IXC
- **Service Type** (Universal Service Order Code): found on CSR with greatest dollars associated with WTN (e.g., 1FB, 1FL, 1MB)
- **Quantity** (number): if WTN is a hunt group pilot and the additional numbers are only associated terminal numbers, then enter total numbers; the monthly recurring charge divided by the quantity equals the per-line charge

CSR Data Entry: Trunk Worksheet

Information from the Trunks section of the LEC CSR will be entered into a Trunk Worksheet.

Fields

- **BTN** (main billing number): account number or telephone number
- **WTN** (working telephone number): telephone number assigned to the trunk group or to individual trunks
- **Service Address** (location): termination of service
- **LEC** (local exchange carrier or local telephone company): incumbent local operating company

Exhibit 18.   Sample Business Line Worksheet

BTN	WTN	Service Address	LEC	Monthly Recurring Charge	LPIC or ZPIC	PIC	Service Type	Quantity
517–337–2445	517–337–2225	2001 Ross Avenue, E Lansing MI 48820	Ameritech	20.07	A13	ATI (= 288)	1MB	1
517–337–2445	517–337–2231	2001 Ross Avenue, E Lansing MI 48820	Ameritech	20.07	A13	ATI (= 288)	1MB	1
517–337–2445	517–337–2531	2001 Ross Avenue, E Lansing MI 48820	Ameritech	20.07	A13	ATI (= 288)	1MB	1
517–337–2445	517–337–2625	2001 Ross Avenue, E Lansing MI 48820	Ameritech	20.07	A13	ATI (= 288)	1MB	1
517–337–2445	517–337–2633	2001 Ross Avenue, E Lansing MI 48820	Ameritech	20.07	A13	ATI (= 288)	1MB	1
517–337–2445	517–337–2741	2001 Ross Avenue, E Lansing MI 48820	Ameritech	20.07	A13	ATI (= 288)	1MB	1
517–337–2445	517–337–2850	2001 Ross Avenue,	Ameritech	20.07	A13	ATI (= 288)	1MB	1

CUSTOMER SERVICE RECORD

3/24/2001	- 22 -	LNNG	1MB	302	666 -337-2445	1
PRINTED	BILLDT	EXCH	CS	CUST	TELNUM	Page

BL GRP	CODE & QNTY	DESCRIPTION		UNIT RATE	TAX TOTAL	FSCMXT
	ZBU	CB, MNF-FMO				
	ZCPI	U				
		MAIN LISTING				
	LN	A*B*C				
	LA	100 Street, City				
	LOC	Suite 1				
	SIC	6141				
		BILLING INFORMATION				
	BN1	ABC Company				
	BA1	100 Street				
	PO	City, State, Zip				
	ZGC	23000000				
	SS	000-00-000Y				
	TAR	NONE				
	ZCPI	U,SYSTEM,9-14-2001				
	1	1MB	PIC ATI/PCA BO, 07-12-96	13.20	13.20	TTNTNN
			ZPIC A13 / LPCA BO, 07-12-96			
			LCC 1M9/HTG A			
	1	SCFXE		0.02	0.02	TTNTNN
	1	TTB		0.00	0.00	NNNNNN
	1	RTV1N		0.00	0.00	TTNTNN
	1	UXWAH		0.58	0.58	NNNNNN
	1	UXTAH		0.24	0.24	TTNTNN
	1	9ZR		5.62	5.62	TTNTNN
	1	NSR		0.41	0.41	TTNTNN
	1	1MB	TN 337-2225/PIC ATI	13.20	13.20	TTNTNN
			PCA BO, 07-12-96/ZPIC A13			
			LPCA BO, 7-12-96/LCC 1M9			
	1	SCFXE	TN 337-2225	0.02	0.02	TTNTNN
	1	TTB	TN 337-2225	0.00	0.00	NNNNNN
	1	RTV1N	TN 337-2225	0.00	0.00	TTNTNN
	1	UXWAH	TN 337-2225	0.58	0.58	NNNNNN
	1	UXTAH	TN 337-2225	0.24	0.24	TTNTNN
	1	9ZR	TN 337-2225	5.62	5.62	TTNTNN
	1	NSR	TN 337-2225	0.41	0.41	TTNTNN
	1	1MB	TN 337-2231 /PIC ATI	13.20	13.20	TTNTNN
			PCA BO, 07-12-96/ZPIC A13			
			LPCA BO, 7-12-96/LCC 1M9			
	1	SCFXE	TN 337-2531	0.02	0.02	TTNTNN
	1	TTB	TN 337-2531	0.00	0.00	NNNNNN

Exhibit 19. Sample Business Line Worksheet Source

- **Monthly Recurring Charge** ($): total monthly billing of all elements associated with the WTN
- **LPIC** *or* **ZPIC** (intraLATA toll provider): in areas where the intraLATA toll provider can be other than the local operating company
- **PIC** (interLATA--ntrastate, interstate, international toll provider): four-digit code indicating the preferred interexchange carrier
- **Service Type** (Universal Service Order Code): found on CSR with greatest dollars associated with WTN (e.g., TFC, T15CX, TFB01, TFU)
- **Quantity** (number): if the WTN is the pilot number in a hunt group and the additional numbers only have terminal numbers associated, then the total numbers need to be entered; the monthly recurring charge divided by the quantity will result in the per-line charge

CUSTOMER SERVICE RECORD

3/24/2001	- 22 -	LNNG	1MB	302	666-337-2445	2
PRINTED	BILLDT	EXCH	CS	CUST	TELNUM	Page

BL GRP	CODE & QNTY	DESCRIPTION	UNIT RATE	TOTAL	TAX FSCMXT
	1	RTV1N TN 337-2531	0.00	0.00	TTNTNN
	1	UXWAH TN 337-2531	0.58	0.58	NNNNNN
	1	UXTAH TN 337-2531	0.24	0.24	TTNTNN
	1	9ZR TN 337-2531	5.62	5.62	TTNTNN
	1	NSR TN 337-2531	0.41	0.41	TTNTNN
	1	1MB TN 337-2625 /PIC ATI	13.20	13.20	TTNTNN
		PCA BO, 07-12-96/ZPIC A13			
		LPCA BO, 7-12-96/LCC 1M9			
	1	SCFXE TN 337-2625	0.02	0.02	TTNTNN
	1	TTB TN 337-2625	0.00	0.00	NNNNNN
	1	RTV1N TN 337-2625	0.00	0.00	TTNTNN
	1	UXWAH TN 337-2625	0.58	0.58	NNNNNN
	1	UXTAH TN 337-2625	0.24	0.24	TTNTNN
	1	9ZR TN 337-2625	5.62	5.62	TTNTNN
	1	NSR TN 337-2625	0.41	0.41	TTNTNN
	1	1MB TN 337-2633 /PIC ATI	13.20	13.20	TTNTNN
		PCA BO, 07-12-96/ZPIC A13			
		LPCA BO, 7-12-96/LCC 1M9			
	1	SCFXE TN 337-2633	0.02	0.02	TTNTNN
	1	TTB TN 337-2633	0.00	0.00	NNNNNN
	1	RTV1N TN 337-2633	0.00	0.00	TTNTNN
	1	UXWAH TN 337-2633	0.58	0.58	NNNNNN
	1	UXTAH TN 337-2633	0.24	0.24	TTNTNN
	1	9ZR TN 337-2633	5.62	5.62	TTNTNN
	1	NSR TN 337-2633	0.41	0.41	TTNTNN
	1	1MB TN 337-2741/PIC ATI	13.20	13.20	TTNTNN
		PCA BO, 07-12-96/ZPIC A13			
		LPCA BO, 7-12-96/LCC 1M9			
		TRAK MODEM			
		TOTAL W/O TAXES		240.60	
		AMT SUBJECT TO FED TAX		222.90	
		AMT SUBJECT TO STATE TAX		222.90	
		AMT SUBJECT TO MUNI TAX		222.90	
	12	NSR Number Portability Charge			
	12	RTV1N 900/976 Call Blocking			
	12	SCFXE AETCP Offset			
	12	UXTAH Ingham Count E911			
	12	TTB TouchTone Service			
	12	UXTAH Ingham Count E911			

Exhibit 19.   Sample Business Line Worksheet Source (Continued)

Sample Trunk Worksheet

See Exhibit 20 for a sample trunk worksheet and Exhibit 21 for its source document.

CSR Data Entry: DID Worksheet

Information from the *Common Equipment* or *System Services* section of the LEC CSR will be entered into a DID Worksheet. The commonly used codes to identify the DID range are ND8, ND9, or NDX. A sample DID Worksheet is listed in Exhibit 22. Its source is shown in Exhibit 23.

Exhibit 20. Sample Trunk Worksheet

BTN	WTN	Service Address	LEC	Monthly Recurring Charge	LPIC or ZPIC	PIC	Service Type	Quantity
817–810–9998	817–810–9998	301 Commerce St., Ft. Worth, TX	SWBT	26.05	9100	222	T15CX	46

		LISTINGS		
	LN	ABC COMPANY		
	LA	100 STREET		
	SA	100 STREET		
		FORT WORTH, TX/DZIP 76201		
	SIC	S8721		
		BILLING INFORMATION		
	BN1	ABC Company		
	BA1	100 Street		
	PO	FORT WORTH, TX 76201		
	TAR	FTW		
	PCL	V		
		COMMON EQUIPMENT		
	ND8	ZTDN 666 670-5500	158.95	158.95
		THRU 5599 /RTI 1301		
	ND9	ZTDN 666 877-2200	13.35	13.35
		THRU 2299 /RTI 1301		
		ZPZAD/SPP VT1	500.01	500.01
		TA 36, 10-13-98		
		9ZRP1	36.72	36.72
		PT8ZX	49.01	49.01
	5	SEHIX	0.26	0.26
		NXN	85.01	85.01
		ZPZAD/SPP VT1	500.01	500.01
		TA 36, 10-13-98		
		9ZRP1	36.71	36.71
		PT8ZX	49.01	49.01
	5	SEHIX	0.26	0.26
		TRUNKS		
	HTML	101-TER HI-48/TLI817		
		810-9998/HTY RG		
810-9998				
	T15CX	TLI 817 810-9998	26.05	26.05
		CFA 110 T1 ZFN 1		
		FTWOTXEDDC5		
		FTWOTXMCMWD2/DES B		
		TGN 1301, D TGN 3 /LHT		
		46 /CLT		
		11.DCJT.817.810		
		9998.T0001		
		LPIC 9100-SBC,050799		
		050799,L		
		PIC 0222-MCI,101598		
		100898,L		
2				
	T15CX	CFA 110 T1 ZFN 1	26.05	26.05
		FTWOTXEDDC5		
		FTWOTXMCMWD2/DES B		
		RTI 1301,TTC 3 /CTL		
		11.DCJT.817.810		
		9998.T0002		
		LPIC 9100-SBC,050799		
		050799,L		
		PIC 0222-MCI,101598		
		100898,L		
3	T15CX	CFA 110 T1 ZFN 1	26.05	26.05
		FTWOTXEDDC5		
		FTWOTXMCMWD2/DES B		
		RTI 1301,TTC 3 /CTL		

Exhibit 21.   Sample Trunk Worksheet Source

		11.DCJT.817.810 9998.T0003 LPIC 9100-SBC,050799 050799,L PIC 0222-MCI,101598 100898,L			
3 4	T15CX	CFA 110 T1 ZFN 1 FTWOTXEDDC5 FTWOTXMCMWD2/DES B RTI 1301,TTC 3 /CTL 11.DCJT.817.810 9998.T0004 LPIC 9100-SBC,050799 050799,L PIC 0222-MCI,101598 100898,L	26.05	26.05
3 5	T15CX	CFA 110 T1 ZFN 1 FTWOTXEDDC5 FTWOTXMCMWD2/DES B RTI 1301,TTC 3 /CTL 11.DCJT.817.810 9998.T0005 LPIC 9100-SBC,050799 050799,L PIC 0222-MCI,101598 100898,L	26.05	26.05

Exhibit 21.   Sample Trunk Worksheet Source (Continued)

Exhibit 22.   Sample DID Worksheet

Area Code	Exchange	Number Start	Number End	Service Address
817	870	5500	5599	301 Commerce St., Ft. Worth, TX
817	877	2200	2299	301 Commerce St., Ft. Worth, TX

Fields

- **Area Code** (three digits): area code
- **Exchange** (three digits): first three digits of a seven-digit telephone number
- **Number Start** (three digits): final four digits of the first DID number in a range
- **Number End** (four digits): final four digits of the last DID number in a range
- **Service Address** (location): physical address, where the service is terminated

CSR Data Entry: Circuit Worksheet

Information from the *Services and Features* section of the CSR will be entered into a Circuit Worksheet. A sample worksheet is shown in Exhibit 24. Its sample source is shown in Exhibit 25.

```
-----------------------------------------------------------------------------------------
|          |        |                 ---LISTINGS---                    |        |        |
|          | LN     | ALCABIADES ENTERPRISES                            |        |        |
|          | LA     | 333 ATTICA WAY                                    |        |        |
|          | SA     | 333 ATTICA WAY                                    |        |        |
|          |        | ATHENS, TX/DZIP 77339                             |        |        |
|          | SIC    | S8721                                             |        |        |
|          |        |                                                   |        |        |
|          |        | BILLING INFORMATION                               |        |        |
|          | BN1    | ALCABIADES ENTERPRISES                            |        |        |
|          | BA1    | 333 ATTICA WAY                                    |        |        |
|          | PO     | ATHENS, TX 77339                                  |        |        |
|          | TAR    | ATH                                               |        |        |
|          | PCL    | V                                                 |        |        |
|          |        |          ---COMMON EQUIPMENT---                   |        |        |
|          | ND8    | /ZTDN 817 870-5500                                | 158.95 | 158.95 |
|          |        | THRU 5599 /RTI 1301                               |        |        |
|          | ND9    | /ZTDN 817 877-2200                                |  13.35 |  13.35 |
|          |        | THRU 2299 /RTI 1301                               |        |        |
|          |        | ZPAZD/SPP VT1                                     | 500.00 | 500.00 |
|          |        | /TA 36, 10-13-98                                  |        |        |
|          |        | 9ZRP1                                             |  36.70 |  36.70 |
|          |        | PT8ZX                                             |  49.01 |  49.01 |
|          | 5      | SEHIX                                             |   0.26 |   1.30 |
|          |        | NXN                                               |  85.00 |  85.00 |
|          |        | ZPAZDISPP VT1                                     | 500.00 | 500.00 |
|          |        | /TA 36, 10-13-98                                  |        |        |
|          |        | 9ZRP1                                             |  36.70 |  36.70 |
|          |        | PT8ZX                                             |  49.01 |  49.01 |
|          | 5      | SEHIX                                             |   0.26 |   1.30 |
|          |        |              ---TRUNKS--                          |        |        |
|          | HTML   | 101-TER HI-48 /TLI 817                            |        |        |
|          |        | 810-9998 / HTY RG                                 |        |        |
| 810-9996 |        |                                                   |        |        |
|          | T15CX  | /TLI 817 810-9998                                 |  26.05 |  26.05 |
|          |        | /CFA 110 T1 ZFN 1                                 |        |        |
|          |        | FTVVOTXEDDC5                                      |        |        |
|          |        | FTVVOTXMCMVYD2 /DES B                             |        |        |
|          |        | TON 1301, D TON 3 /LHT                            |        |        |
|          |        | 46 /CLT                                           |        |        |
|          |        | 11.DCJT.817.810.                                  |        |        |
|          |        | 9998.T0001                                        |        |        |
|          |        | /LPIC 9100-SBC, 050799                            |        |        |
|          |        | 050799,L                                          |        |        |
|          |        | /PIC 0222-MCI, 101598                             |        |        |
|          |        | 100898, L                                         |        |        |
-----------------------------------------------------------------------------------------
```

Exhibit 23.   Sample DID Worksheet Source

Fields

- **BTN** (main billing number): account number or telephone number
- **Circuit Number:** unique alpha/numeric code
- **Monthly Recurring Charge** ($): total monthly billing of all elements associated with the circuit
- **Service Address 1** (location): first point of termination of service
- **Service Address 2** (location): second point of termination of service
- **Purpose** (TIE line, carrier services, DCS circuit): firm provided information
- **Use:** voice, data, etc.
- **Carrier** (LEC): local provider name
- **Reference:** corresponding circuit numbers

Exhibit 24.   Sample Circuit Worksheet

BTN	Circuit Number	Monthly Recurring Charge	Service Address 1	Service Address 2	Purpose	Use	Carrier	Reference
210-377-5660	14.HCGF.525077	307.70	301 Commerce St., Ft. Worth, TX	2001 Ross Ave, Dallas, TX	DCS	Voice	SWBT	NA

```
| 510 066 3140 749 |                   |                                          |   |   |
|                  |                   |                                          |   |   |
|                  |                   |      --- ACCOUNT IDENTIFICATION ---       |   |   |
|                  |                   |                                          |   |   |
|                  | BILLED TO         | ABC COMPANY                              |   |   |
|                  |                   | 100 COMMERCE STREET                      |   |   |
|                  |                   | FORT WORTH, TX/DZIP 76201                |   |   |
|                  |                   |                                          |   |   |
|                  | CUSTOMER S SERVICE ADDRESS                                   |   |   |
|                  |                   |     100 COMMERCE STREET                  |   |   |
|                  |                   |                                          |   |   |
|                  |                   |      --- SERVICES AND FEATURES ---        |   |   |
|                  | CLS               | 14.HCFG.525077                           |   |   |
|                  |                   | /NC HCHK                                 |   |   |
|                  |                   | /PIU 100                                 |   |   |
| 1                | XDH1X             |                                          |   |   |
|                  | CKL               | 100 COMMERCE STREET, FT WORTH,           |   |   |
|                  |                   | TX/DES LOC SUITE 308/SN                  |   |   |
|                  |                   | ABC COMPANY                              |   |   |
|                  |                   | 04DS9.1S/LSCO 827 810/ACTL1              |   |   |
| 24               | S25EX             |                                          |   |   |
|                  | TEMCS             | /SPP TP10-06-94 1 60                     |   |   |
|                  |                   | /RZN 01                                  |   |   |
|                  |                   | INTER 100%                               |   |   |
|                  |                   | ( 108.0 x 1 )                            | $180.00 | |
| 11               | 1L5XX             | SPP TP06094 1 60                         |   |   |
|                  |                   | RZN 01                                   |   |   |
|                  |                   | INTER 100%                               |   |   |
|                  |                   | ( 37.5 + ( 11.20 x 11 ))                 | $160.70 | |
| 1                | R18AB             |                                          |   |   |
| 1                | MUF               | /EUTA AND                                |   |   |
| 1                | CKL               | 2-2000 XX AVE., DALLAS,                  |   |   |
|                  |                   | TX/LOC FLR 1800/SN BELL                  |   |   |
|                  |                   | DLLSTXROK01/LSO 214                      |   |   |
| 24               | S25EX             |                                          |   |   |
| 1                | R18AB             |                                          |   |   |
| 1                | PT6               |                                          |   |   |
|                  |                   | INTER 100%                               |   |   |
|                  |                   | ( 39.00 x 1 )                            | $39.00 |  |
|                  |                   | NON PAYMENT PLAN INTERSTATE SUBTOTAL     | $39.00 |  |
|                  |                   |   PAYMENT PLAN INTERSTATE SUBTOTAL       | $268.70 | |
|                  |                   |                                          |   |   |
|                  |                   | CIRCUIT TOTAL                            | $307.70 | |
```

Exhibit 25. Sample Circuit Worksheet Source

2.2 Enter LEC Invoice Information

Materials needed

The LEC bill is needed for this procedure. The LEC bill includes a summarization of the monthly recurring chargtes and itemized call. The call detail includes the number dialed, duration, and charge.

Enter billing information.

Follow these steps to enter billing information from the LEC bill.

Exhibit 26.   Sample LEC Charges Worksheet

BTN	WTN	Amount	Duration	Call Type
214-953-0669	214-953-0669	5.30	26	IntraLATA toll

1. Review the bill and note all WTNs with usage.
2. Enter the WTN into the database.
3. Enter "X" in **LEC Bill Source** field to indicate that the data was obtained from the LEC bill.
4. Enter the usage charge associated with the number.
5. Enter the billed telephone number (BTN).
6. Enter the service type in the **Service Type** field (e.g., 1MB, 1FB, T1, or ISDN).
7. Is there a service address in the database for the number?
8. Add new service address into the database.
9. Enter LEC Bill in the **Address Source** field.
10. Select the appropriate service address from the **Service Address** field.
11. Initial the hardcopy LEC bills.

Worksheet

Information from the LEC bill will populate one of two different project worksheets:

- **LEC Charges** (itemized calls): page will have LEC logo
- **IXC Charges (on LEC Bill)** (itemized calls): page will have IXC logo

LEC Data Entry: LEC Charges Worksheet

The LEC bill has itemized call detail. Usage-sensitive charges are recorded by call type: local measured service, intraLATA toll, collect, and credit card charges. Information from this section will be entered into a LEC Charges Worksheet (see Exhibits 26 and 27).

The **Amount** field should be populated with a $0 for all WTNs associated with a BTN if no usage was found on the corresponding LEC bill.

Fields

- **BTN** (main billing number): account number or telephone number
- **WTN:** Specific Number
- **Amount** ($): total billing of all calls associated with a number for a specific type of call
- **Duration** (minutes): total duration for all calls associated with a number for a specific type of call
- Call Type

Southwestern Bell

Account Number		
972-419-2000-444-5		
July 29, 1999		

Billing for:	**Current Charges**	
SWBell Telephone	Monthly Service - Jul 29 thru Aug 28	26.55
	Municipal Charge	2.61
	Expanded Local Calling Service Surcharge	0.26
	Number Portability Service Charge	0.33
	FCC Approved Customer Line Charge	7.34
	Amount Subject to Sales Tax: 37.09	
	911 Service Fee	1.52
	Texas 911 Equalization Surcharge	0.02
	Texas Poison Control Surcharge	0.02
	TX USF Charge 1.52%	0.60
	Other Charges (See items 1 and 2)	0.02
	Itemized Calls (See items 3 thru 12)	5.30
	SWBell Telephone Current Charges (before taxes)	44.57
	Federal Tax	1.29
For Billing Questions:	State and Local Tax	3.55
214-571-7768		

Other Charges	Item	Explanation			Amount
	1	Rate changed on Jul 01	From	To	
		- FCC Approved Customer Line Charges	7.19	7.34	
			7.19	7.34	
		Monthly Rate			
		Charge for change in rates Jul 06 thru July 28			0.14
	2	Rate changed on July 06	From	To	
		Number Portability Service Charge	0.48	0.33	
			0.48	0.33	
		Monthly Rate			
		Credit for change in rates Jul 06 thru July 28			0.12 CR
		SWBell Telephone Total Other Charges (before taxes)			0.02

Itemized Calls	Item	Date	Time	Place Called		Area	Number	Rate*	Min	Amount
		Calls from 214-953-0669								
	3	07/02	12:02PM	FORT WORTH	TX	817	877-2260	D	3	0.73
	4	07/02	01:47PM	FORT WORTH	TX	817	877-2260	D	1	0.25
	5	07/13	03:41PM	FORT WORTH	TX	817	877-2260	D	1	0.25
	6	07/13	04:52PM	ARLINGTON	TX	817	607-1575	D	5	0.63
		Total Itemized Calls for SWBell Telephone (before taxes)							10	1.86

For Your Information	On an annual basis, Southwestern Bell is required to file information with the Federal Communications Commission (FCC) regarding costs for providing customers with access to long distance networks. Analysis of this information and gradual implementation of FCC rule changes produce rate fluctuations. As a result of our 1999 Annual FCC filing, the Federal Prescribed Interexchange Carrier Charge (PICC) has increased from 53 cents per month $1.04 per month for each **primary** residential phone line and business line. The PICCs for **additional** phone lines at residences and businesses have decreased substantially from $1.50 to 57 cents per line for residences and form $2.21 to 57 cents per line for businesses. **These changes became effective July 1, 1999, and only affect the Southwestern Bell portion of phone bills for those customers who have not chosen a long distance carrier for 1+ dialing from their phone lines.**

Exhibit 27.   Sample LEC Charges Worksheet Source

LEC Data Entry: IXC Charges Worksheet

The LEC bill includes IXC charges, normally itemized call detail. Usage-sensitive charges are recorded by call type: toll, collect, and credit card charges. Information from this section will be entered into an IXC Charges Worksheet. Exhibits 28, 30, and 32 are sample worksheets. Exhibits 29, 31, and 33 are their corresponding source documents.

Exhibit 28. IXC Charges Worksheet Sample 1

LEC	IXC	BTN	WTN	Amount	Duration	Call Type or Charge Description
SWBT	AT&T	214-419-2900	214-419-2345	4.51	7	Collect

Exhibit 29. IXC Charges Worksheet Source Sample 1

AT&T

Account Number
972-419-2000-444-5
July 29, 1999

Billing for:	**Current Charges**					
AT&T	Itemized Calls (See item 1)					4.51
						4.51
	AT&T Current Charges (before taxes)					
	Federal Tax				0.14	
For Billing Questions:	State and Local Tax				0.00	
1-800-325-0138						

Itemized Calls	**Item**	**Date**	**Time**	**Place Called**		**Area**	**Number**	**Rate***	**Min**	**Amount**
		Calls from 972-419-2345								
	1	07/28	03:37PM	ADDISON	TX	972	419-2345	DS	7	
				COLLECT FROM MINNEAP	MN	612	964-7471			4.51
				Total Itemized Calls for AT&T (before taxes)					7	4.51

For Your Information MOVING? ADDING LOCATIONS? NEED ADVICE ON LONG DISTANCE SERVICE? CALL AT&T ON 1 800 222-0400

Exhibit 30. IXC Charges Worksheet Sample 2

LEC	IXC	BTN	WTN	Amount	Duration	Call Type or Charge Description
SWBT	American Tel	214-999-1400	214-999-1234	65.19	53	Dialed direct
SWBT	American Tel	214-999-1400	214-999-1234	5.00	NA	Monthly charge

AMERICAN TEL

Account Number   SERVICES
214-999-1400-999
July 29, 1999

Billing for: **American Tel**	**Current Charges** Monthly Charge Itemized Calls (See items 1 to 4)							5.00 65.19 70.19

Network Operator Current Charges (before taxes)

American Tel   Federal Tax   1.95
For Billing Questions   State and Local Tax   0.79
1-800-530-4898

Itemized Calls	**Item**	**Date**	**Time**	**Place Called**		**Area**	**Number**	**Rate***	**Min**	**Amount**
		Calls from 214-999-1234								
	1	07/04	02:51PM	MINNEAP	MN	612	964-7471	DS	12	14.76
	2	07/07	03:37PM	HOUSTON	TX	713	999-9999	DS	8	9.84
	3	07/10	09:43AM	LA	CA	310	888-7766	DS	9	11.07
	4	07/09	05:22PM	QUEENS	NY	516	444-1256	DS	24	29.52
		Total Itemized Calls for American Tel (before taxes)							53	65.19

Exhibit 31.   IXC Charges Worksheet Source Sample 2

Exhibit 32.   IXC Charges Worksheet Sample 3

LEC	*IXC*	*BTN*	*WTN*	*Amount*	*Duration*	*Call Type or Charge Description*
SWBT	BPSI for YP NET	214-999-1400	214-987-2882	25.14	NA	Monthly charge

BPSI

Account Number
214-999-1400-999
July 29, 1999

Billing for: **YP Net**	**Current Charges** Monthly Charge				25.14 25.14

Network Operator Current Charges (before taxes)

BPSI   Federal Tax   1.01
For Billing Questions   State and Local Tax   0.75
1-800-800-9999

Charges billed to 214-987-2882

Exhibit 33.   IXC Charges Worksheet Source Sample 3

There may be more that one IXC with charges billed on the LEC bill. Dialed direct calls billed by an IXC on a LEC bill may be an indication that one or more lines have been "slammed."

There may be monthly recurring charges from the IXC on the LEC bill. These charges may be unauthorized and represent "cramming."

Fields

- **LEC** (company name of billed): LEC name on bill
- **IXC** (company name or logo on top of specific pages): IXC
- **BTN** (main billing number): account number or telephone number
- **WTN** (specific number): WTN, DID number, or calling card number
- **Amount** ($): total billing of all calls associated with a number for a specific call type
- **Duration** (minutes): total duration for all calls associated with a number for a specific call type
- **Call Type** *or* **Other Charge Description** (DDD, OD, Calling Card, Yellow Page Charges): cramming charges noted in this field

2.3 Enter IXC Billing Information

The IXC bill includes monthly recurring charges and/or itemized call detail. Itemized call detail includes number dialed, duration, and charge.

Although there are a number of IXCs, the main ones are MCI and AT&T. A firm may use more than one carrier but should have a contract with the primary carrier. The primary carrier should be identified along with the product and/or services subscribed to. Special attention should be focused on carrier bills that are not part of subscribed or contracted services.

Using CD-ROM Bills

The IXC (interexchange carrier) bill can provide invoice, call detail, and other information in electronic form (see Exhibit 34). The predominant form is on a CD-ROM.

The IXC CD-ROM contains several files that have information necessary to properly complete the audit. Not all files need to be accessed. Not all fields of the accessed files need to be extracted.

Paper Bill Worksheets

Information from the IXC paper bills will populate one or more of several worksheets (see Exhibit 35).

Exhibit 34. Carrier Services

Carrier Name	Contract	Offering	Description	CD-ROM
AT&T	Y	VTNS, SDN, SDNOne Net, custom contract	Bundled offerings, which include SDN (1+ long distance, and calling cards), Megacom (800 dedicated), Readyline (800 switched), Frame Relay	Y
MCI	Y	VNET	1+ and calling cards	Y
AT&T	N	Readyline	Toll-free switched	N
AT&T	N	Small business bills	1+, calling cards, collect, 900 calls, operator-assisted calling	N
MCI	Y	800 Perspective	Toll-free switched and dedicated	Y
AT&T	N	Megacom	Toll-free dedicated	Y
AT&T	Y	Private line	Facilities, circuits	Y

Exhibit 35. IXC Paper Bill Information

Worksheet Name	Description	Source	CD-ROM
Principal IXC 1+	Summary of 1+ use by rate schedule or call type	SDN, VTNS, VNET	Y
Secondary IXC 1+	Summary of 1+ use by rate schedule or call type	SDN, VTNS, VNET	Y
Non-Contract IXC 1+	Summary of 1+ charges	Small Business Bills	N
Principal IXC Toll-free	Summary of toll-free use by type	VTNS, SDNOneNet, 800 Perspective	Y
Secondary IXC Toll-free	Summary of toll-free use by type	VTNS, SDNOneNet, 800 Perspective	Y
Non-Contract IXC Toll Free	Summary of toll-free charges	Readyline, Megacom	N
Principal IXC Calling Card	Summary of calling card use by rate schedule or call type	SDN, VTNS, SDNOneNet	Y
Secondary Calling Card	Summary of calling card use by rate schedule or call type	SDN, VTNS, SDNOneNet	Y
Non-Contract Calling Card	Summary of calling card use by call type		Y
Facilities	Circuit specifics and recurring charges	Private Line Bills, VTNS, SDN	Y
PICed Lines	Lines provisioned by IXC in client or contract specific database	File specific	Y
Toll-Free Numbers	Listing of toll-free numbers and termination numbers	File specific or query of call detail file	Y

Exhibit 36. IXC CD-ROM Information

Worksheet	Description	Type of Bill
IXC 1+	Summary of 1+ use by rate schedule or call type	SDN, VTNS, VNET
IXC Toll Free	Summary of toll-free use by type	VTNS, SDNOneNet, 800 Perspective
Contract Calling Card	Summary of calling card use by rate schedule or call type	SDN, VTNS, SDNOneNet
Circuits	Circuit specifics and recurring charges	Private line bills, VTNS, SDN
PIC Charges	All lines provisioned in the client specific portion of the IXC database	

CD-ROM worksheets

Information from the IXC CD-ROM will populate one or more of several worksheets (see Exhibit 36).

IXC Data Entry: Principal (and Secondary) IXC 1+ Paper Bill Worksheet

The primary interexchange carrier (IXC) used by the firm. The contracted IXC that supplies 1+ switched or dedicated service (see Exhibit 37 for a sample IXC Charges Worksheet and Exhibit 38 for its source).

Worksheet

Information from the IXC bill will populate a worksheet containing the following fields.

- Carrier Name
- **Account Number** (main billing number)
- Bill Group
- **WTN** (specific number)
- **Amount** ($): total billing (include discount) of all calls associated with number
- **Duration** (minutes): total duration for all calls associated with a number for a specific type of call
- **Call Type**: intraLATA-local, intraLATA-toll, interstate
- **Origination Type:** switched, dedicated
- **Termination Type:** switched, dedicated

Steps

To populate this worksheet's example, make the following calculations. The objective is to populate intrastate and interstate post-discounted spends.

Exhibit 37. Sample IXC Charges Worksheet

Carrier Name	Account Number	Bill Group	WTN	Amount	Duration	Call Type	Org Type	Term Type
AT&T SDN OneNet	1000-262-0481	483	817 332-2243	$596.61	5265.8	All	Switched	All
OR								
AT&T SDN OneNet	1000-262-0481	483	817 332-2243	$192.13	1777.5	Intrastate	Switched	All
AT&T SDN OneNet	1000-262-0481	483	817 332-2243	$315.54	384.2	Interstate	Switched	All
AT&T SDN OneNet	1000-262-0481	483	817 332-2243	$81.74	104.1	International	Switched	All
OR								
AT&T SDN OneNet	1000-262-0481	483		$596.61	5265.8	All	Switched	All

AT&T OneNet Service					SUMMARY OF INVOICE CHARGES	
ABC COMPANY				**Account Number:**		**1000-262-0481**
				Invoice Number:		2104 102619
				Invoice Date:		**9/1/1998**
				For Billing Inquiries:		1-800-989-0130
Item No.	Service Description	Usage Charges	Monthly Charges	One-Time and Prorated Charges/Credits	Taxes, Fees and Surcharges	Total
Discounts						
1	Term and Column Plan (SDN Domestic)			$415.11 CR		$415.11 CR
2	Term and Column Plan (SDN International)			$30.35 CR		$30.35 CR
3	International Term Plan			$14.89 CR		$14.89 CR
	Total Discounts:	$0.00	$0.00	$460.35 CR		$460.35 CR
Discounts						
Direct Dial						
4	Intrastate	$349.32			$23.51	$372.83
	455 Calls 1,777:3:0 Mins:Secs					
5	Interstate	$573.70			$33.52	$607.22
	883 Calls 3,384:12:0 Mins:Secs					
6	International	$119.14			$7.84	$126.98
	15 Calls 104:06:0 Mins:Secs					
7	Directory Assistance	$14.80			$1.67	$16.47
	15 Calls					
	Total Direct Dial:	$1,056.96	$0.00	$0.00	$66.54	$1,123.50
	Total Usage:	$1,056.96	$0.00	$0.00	$76.05	$1,366.95
Features						
Network						
8	Access Line Grouping		$0.00	$0.00		$0.00
	Total Network Features:	$0.00	$0.00	$0.00	$0.00	$0.00
Regulatory Charges						
9	Access Line Grouping		$0.00	$47.50	$3.60	$51.10
10	Universal Conductivity		$0.00	$18.31	$1.39	$19.70
	Regulatory Charges:	$0.00	$0.00	$65.81	$4.99	$70.80
	Total:	$1,056.96	$0.00	$394.54 CR	$71.53	$733.95
	Total Usage:	$1,056.96	$0.00	$0.00	$66.54	$1,123.60

Exhibit 38.   Sample IXC Charges Worksheet Source (continued on next page)

1. Discount rate = Discount dollars/Pre-discounted intrastate amount + Pre-discounted interstate. Example: $460.3539/($349.32 + $573.70) = 0.50
2. Calculate the post-discount spend. Example: Intrastate $349.32 * 0.55 = $192.13; Interstate $573.70 0.55 = $315.54
3. Calculate the post-discount international spend. Example: International spend – International discounts: $126.98 – $30.35 – $14.89 = $81.74

IXC Data Entry: Principal (and Secondary) Toll-Free Services Worksheet

The Toll-Free Services Worksheet will be populated manually from paper bills or mechanically from an IXC electronic medium such as a CD-ROM (see Exhibit 39 for a sample worksheet).

Bill Group	BTN/ Location Name	Number of Calls	Duration Min:Sec: Tenths	Pre-Discount Usage Charges	Access and Features Charges	Discounts Credit Amount	Taxes/ Fees	Total Post-Discounted Charges	Average Cost per Minute*
483 Acct Level		0	0:00:0	0	$65.81	$0.0000	$4.99	$70.80	$0.0000
	817-332-2243	1368	5,265:48:0	$1,056.96	$0.00	$460.3539	$66.54	$663.14	$0.1105
Bill Group Totals:		1368	5,265:48:0	$1056.96	$65.81	$460.3539	$71.53	$733.94	$0.1105

AT&T OneNet Service

ABC COMPANY

LOCATION SUMMARY REPORT

Bill Group Summary

Account Number: **1000-262-0481**

Invoice Number: 2104 102619

Invoice Date: **9/1/1998**

For Billing Inquiries: 1-800-989-0130

* Average Cost per minute is pre-Discounted Charges minus Discounts Divided by Duration Directory Assistance, Monthly, One-Time and Pro-Rated Charges, Taxes, Surcharges and fees included

Local Feature Charges, Local line Charges and Local Usage Charges are not included in this report

Exhibit 38.   Sample IXC Charges Worksheet Source (Continued)

Exhibit 39.   Sample Worksheet

800 Number	Termination Number	Usage	Type of Usage	Duration	IXC	Termination Type
8008732678	9726616550	53.43	Interstate	439	MCI	Switched
8008732678	9726616550	84.30	Intrastate	581	MCI	Switched
8882705741	8070	178.03	Interstate	2231	MCI	Dedicated
8882705741	8070	809.92	Intrastate	9865	MCI	Dedicated
8008732678	9726616550	957.06	Interstate	7781	ATT/OneNet	Switched
8008732678	9726616550	1160.57	Intrastate	8231	ATT/OneNet	Switched
8008732678	9726616550	294.73	Interstate	4399	ATT/OneNet	Dedicated
8008732678	9726616550	372.82	Intrastate	5178	ATT/OneNet	Dedicated
8008732678	9726616550	1167.15	Interstate	7781	ATT/Readyline	Switched
8008732678	9726616550	746.64	Intrastate	4392	ATT/Readyline	Switched

Before You Begin

Obtain IXC bills with the toll-free usage for a site. Paper copies of invoices may provide all detail necessary to populate the worksheet.

The Toll-Free Services Worksheet is a summary document. The summary is the total of the minutes and duration associated with a toll-free number and a unique termination in one month.

Worksheet

Follow these steps to enter the toll-free usage information from the IXC bill.

- Toll-Free Number
- Termination Number
- **Usage:** dollar amount
- **Type of Usage:** intrastate, interstate, other
- **Duration:** minutes; seconds
- **IXC:** carrier and carrier service type (source of the information)
- **Termination Type:** switched or dedicated

IXC Data Entry: PICCed Lines Worksheet

The Preferred Interexchange Carrier Charge is referred to as the PICC. Each line from a local exchange company provisioned on an IXC network generates a PICC. Most carriers bill the PICC as a separate element. Some IXCs include the PICC in the usage charges.

Purpose

The PICCed Worksheet is used to verify the existence of all switched lines and the chosen IXC per line. Lines provisioned in the primary IXC's database, which is client specific, can be extracted from the CD-ROM.

Each switched line generates a PIC fee. A review of the switched services provisioned with the primary IXC is part of the analysis process.

Worksheet

- **Account Number:** IXC billing number
- **SALES_OFF** (the AT&T office that handles this account): can be deleted
- **MCN** (master account number; client-specific code): can be deleted
- **BILL_GRP** (Bill Group): field may not be populated; the BTN can be used to determine the BILL_GRP by referencing the Acctsum.dbf
- **BTN:** billing telephone number
- **WTN:** working telephone number associated with PIC
- **LINE_TYPE:** the type of line service associated with the WTN: M = multiple business line; C = Centrex; cellular; ISDN
- **STATE_CODE:** state in which the line is terminated
- **PICC_CHARGE:** actual dollar amount for PIC fee (monthly recurring charge): M = $2.50; C = .31; S = .53 (Illinois); B = 1.50.

Exhibit 40 shows a sample of an AT&T PIC file imported from the CD-ROM to an Excel spreadsheet.

Extracting the AT&T PICC file from the Billing Edge CD-ROM

1. *Open* — Excel
2. *Open* — File Open

Exhibit 40.   Sample AT&T PIC File

SALES_ OFF	MCN	BILL_ GRP	BTN	WTN	LINE_ TYPE	STATE_ CODE	PICC_ CHRG
	10002619715		2032915883	2032915884	M	ZZ	2.50
	10002619715		5163278800	5163278801	C	NY	.31
	10002619715		5163278800	5163278802	C	NY	.31
	10002619715		5163278800	5163278803	C	NY	.31
	10002619715		5163278800	5163278804	C	NY	.31
	10002619715		5163278800	5163278805	C	NY	.31

3. *Open* — D: (drive where the CD-ROM is located.)
4. *Open Folder* — Sdninv
5. *Change Files of Type* — dBase Files (*.dbf)
6. *Open File* — WTNINVCL.dbf

This file contains all switched lines provisioned on client service. Each line generates a PICC fee. The fee is based on line type.

Note that MS Access or some other database program can be used if the detail records exceed the capacity of Excel.

IXC Data Entry: Facilities Worksheet

The Facilities Worksheet will be populated manually from paper bills or mechanically from an IXC electronic medium such as a CD-ROM.

Purpose

The Facilities Worksheet is used to verify the existence of dedicated T1 facilities to the IXC's point of presence (POP). Knowledge of the existing services provisioned on the T1 will be of assistance in the analysis.

Fields

- STATE
- CITY
- STREET (street address): termination of T1
- **PHONE** (BTN): number used to identify traffic
- **CIRCUIT NUMBER** (Voice Circuit IDs): carrier-specific circuit number
- **MONTHLY RECURRING CHARGES** (Monthly Recurring Access): monthly charges
- **ACCOUNT NUMBER:** carrier-specific circuit number
- **INBOUND/OUTBOUND:** types of origination and termination

- **SERVICE TYPE:** toll-free, inbound on Net, etc.
- **CHANNEL NUMBER:** if T1 is channeled into more that one trunk group or if T1 is part of a larger trunk group

Steps

1. *Open* — Access
2. *Open* — Blank Database
3. *Create name for new database* — suggestions: firm name or project name, or default to db3.mdb
4. Open file
5. *Get external data* — import from CD
6. *Go to* — location of CD and corresponding files
7. *Change "Files of Type"* — for example, to dBase III (*.dbf)
8. Go to CD-ROM location
9. *Open folder* — SDN
10. *Import* — CIRINV.dbf
11. Open CIRINV.DBF
12. *Delete all columns except* — IB_SUBACCT Field, BTN Field, DIALED_Toll Free Field
13. *Query*: creation
14. Select SUM8USG
15. *Select Fields*: all
16. Finish building query
17. *Qualify Query*: make M0_AMT>0
18. Check results
19. *Export:* create Excel spreadsheet

The resulting Excel file will contain numbered circuit types (see Exhibits 41 and 42). The AT&T Voice T1 is identified with circuit numbers that begin with DHEC. T1 may or may not be provisioned for voice. T1 may be channelized with a subset of the 24 channels used for voice. AT&T T1 may be billed by another product (private line bills) and may not be on the CD-ROM. MCI T1 and other circuit types are all billed on paper.

2.4 Enter Firm Information

Before beginning

Obtain the following lists:

- Working telephone numbers and associated service addresses
- Circuit numbers and service addresses
- Equipment type and service addresses
- LEC contacts and phone numbers
- PBX reports

Exhibit 41. Typical MCI T1 Listing

State	City	Street	Phone	Voice Circuit ID	Monthly Recurring Access	Account Number	Inbound/ Outbound	Service Type	Channel Number
CA	Century City	1880 Century Park	310 201 1847	IBW092290001–24	$300.00	N1998449	Outbound	Vnet	01–24
CA	Costa Mesa	575 Anton Rd	714 435 8600	IBW778940001–24	$300.00	N1989030	2-Way	Vnet	01–24
CA	Los Angeles	400 South Hope Street	213 236 3000	VAD556080001–5	$300.00	N0279695	Outbound	Vnet	00
CA	Los Angeles	400 S Hope St	213 236 3000	IBX652030001–24	$300.00	N2062464	Outbound	Vnet	01–24
CA	Menlo Park	68 Willow Rd	650 322 0606	IBBBGC7H0009–24	$300.00	N2348364	Outbound	Vnet	09–24
CA	Palo Alto	525 University Ave	650 329 3485	IBBBH8RW0001–20	$300.00	N2387598	2-Way	Vnet	01–20
CA	San Diego	750 B St	619 231 1200	VAM197270001–20	$300.00	N2926683	Outbound	Vnet	01–24

Exhibit 42.   Typical AT&T T1 Listing

State	City	Address	BTN	Circuit No	Monthly Recurring Charges
TX	Dallas	5BC Pericles Way	214–754–7900	DHEC.366152.1.ATI	$673.82
TX	Dallas	5BC Pericles Way	214–754–7900	DHEC.439916.100	$1066.55
TX	Dallas	5BC Pericles Way	214–754–7900	DHEC.453506.ATI	$603.76
TX	Fort Worth	1 Democracy Street	817–332–2243	DHEC.814727.ATI	$616.98
CA	San Francisco	2 Hoplite Drive	415–957–3000	DHEC.419166.ATI	$444.96
CA	San Francisco	2 Hoplite Drive	415–957–3000	DHEC.494783.ATI	$544.80
CA	San Francisco	2 Hoplite Drive	415–957–3000	DHEC.537538.100	$960.80

Enter company information

Follow these steps to enter the company information.

1. Sort the material for the site.
2. Enter the location name, address, and billing telephone number (BTN) into the database.
3. Enter the local exchange carrier (LEC) name.
4. Enter the equipment type.
5. Enter the working telephone numbers (WTNs) for the site.
6. Enter the circuit numbers for the site.

Data Entry: PBX Report Trunk Information

Before beginning

Complete the Request Switch Reports procedure before beginning this procedure.

When to use

Use this procedure after entering the IXC and LEC WTN information.

Enter trunk information

Follow these steps to enter the trunk information for an Avaya switch.

1. Sort the **Service Type** field by ascending order.
2. Create a new worksheet.
3. Copy and paste the data from the following columns into the new worksheet:
   a. WTN/Circuit Number
   b. Service Type
   c. Total

No.	Existence client Info	Existence LEC CSR	Existence LEC Bill	Existence IXC Bill	Action	Notes
1	^				A, B, C	Update company records.
2	^		^		B, C	Need the CSR to identify the IXC.
3	^		^	^	B	
4	^			^	B	
5	^	^	^		B	Bill exists, no usage Bill doesn t exist, request LEC bill for the BTN
					C	
6	^	^	^	^	C	
7	^	^		^	B	Bill exists, no usage Bill doesn t exist, request LEC bill for the BTN (non working number)
8		^			B	Bill exists, no usage Bill doesn t exist, request LEC bill for the BTN
				^	C	Individual at the site installed line, PICed to another IXC, no usage, perhaps an authorized service request.
9		^	^	^	B	
10		^	^		C	
11		^	^	^	D	
12			^		C	
13			^	^	B	
14				^	A, B, C	If no usage then ABC, if usage, find the LEC. Probably have an incorrect LEC and check for the correct LEC.
15	^	^	^	^	E	

A = Verify the number is no longer in service.
- Dial the number
- Listen for message
- Indicate the line is disconnected

B = Follow-up with the LEC.
- Verify with the LERG (Local Exchange Routing Guide) or the client information that the correct LEC is identified for the WTN.
- Confirm the LEC CSR and bill request date.
- Contact the LEC and request document re-send.

C = Follow-up with the IXC.
- Verify with the client information that the correct IXC is identified for the WTN, and/or check CSR for the PIC.
- Confirm the IXC bill request date.
- Contact the IXC and request document re-send.

D = Update client
- Inform client that the phone number or circuit information needs to be updated.

E = Proceed to next step

Exhibit 43. Completeness of Documentation

4. Create the following column headers:
   a. Switch
   b. Trunk Group Number
   c. Trunk Group Member Number
   d. Trunk Status

2.5 Verify Completeness of Documentation

Without information from all four sources — the firm, LEC bill, LEC CSR, and the IXC bill — savings or recovery opportunities may be significantly reduced.
You should consider the items in Exhibit 43 when obtaining the CSRs.

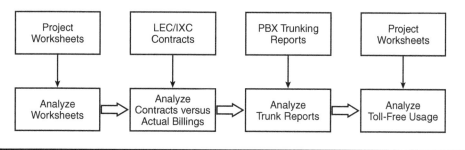

Exhibit 44. Analyzing the Data

3. Analyzing the Data

The Analyze Data phase reviews data for recovery and savings opportunities.

Flowchart

The flowchart in Exhibit 44 shows the major steps involved in this phase and the external entities that provide relevant information.

Analyze data

The Analyze Data phase includes the following:

3.1 *Analyze worksheets.* This identifies findings in the various worksheets.
3.2 *Analyze contract rates versus actual billed rates.* This validates carrier billing.
3.3 *Examine trunk reports.* Examine trunk reports for trunk termination.
3.4 *Analyze toll-free usage.* Analyze toll-free usage for low activity.

3.1 Analyze Worksheets

Worksheets developed during Phase 2 are analyzed separately and compared against each other to determine errors and cost savings.

Single Worksheet Analysis

- *CSR Business Lines*: review of the quantity, use, and purpose of all lines that are not terminated in a PBX
- *LEC Charges*: review of the usage sensitive charges associated with business lines
- *LEC IXC Charges*: review of types of charges and total spend per line
- *Principal IXC Usage, Contract versus Actual*: compares contract rates with actual charges for all usage-sensitive calls by service and type
- *Principal IXC Facilities, Tariff versus Actual*: compares contracted or lowest tariff rate with actual billed rate

Multi-Worksheet Analysis

- *Switched versus Dedicated* (1+). Include some or all of the following worksheets: Trunk, Trunk CSR, Principal IXC 1+ Use, Secondary IXC 1+ Use, LEC Charges, and IXC Charges (LEC bills).
- *Switched versus Dedicated* (Toll-Free). Include some or all of the following worksheets: Trunk, DID, Principal IXC Toll-Free + Use, and Secondary IXC Toll-Free + Use.
- *Calling Cards.* Review calling card use, including the use of the DID Worksheet.

Worksheet Analysis: CSR Business Line Charges

Business line charges are the first to be reviewed (see Exhibit 45). Information is obtained from the CSR.

Worksheet analysis steps

1. Merge information from Business Line Charges Worksheet with company inventory.
2. Request LEC information for lines not previously requested.
3. Extract lines with addresses not identified. If there are usage charges associated with the line, further investigation may be required.
4. For numbers not identified, dial the number. Make note if number answers with modem or fax tone, does not answer, receives a busy signal, or belongs to another company.
5. Compare service address with known service addresses. For lines with service addresses not known that have an unknown purpose, recommend disconnect.
6. Classify all remaining lines that are unidentified as recommended for disconnection. If the quantity of lines exceeds ten and a PBX serves the location, a site visit may be in order.
7. Evaluate monthly recurring charges by comparing them to other charges for the same type of service (e.g., if the monthly recurring charge for one 1MB is $20.07 and for another is $32.02). Refer to the actual CSR to determine the type of charges that made the one item higher than the other.
8. Determine whether the LPIC can be changed. Look for the USOC (Universal Service Order Code) on the CSR Service Type with the highest WTN charges. The LPIC is their carrier of choice for intraLATA toll when dialing 1+ the number. The LEC puts an LPIC on a CSR even in locations that do not use LPIC functions. IntraLATA toll using the principal carrier can be accomplished by dialing 101 + PIC Access Code + 1 + Number.
9. Evaluate the total number of lines given the size of the location or the number of employees.

Examples

Exhibit 46 contains sample findings from the Business Line Worksheet.

Exhibit 45. Sample Business Line Charges Worksheet

Item	BTN	WTN	Service Address	LEC	Monthly Recurring Charge	LPIC or ZPIC	PIC	Service Type	Quantity
1	517–337–2445	517–337–2225	2001 Ross Avenue, E Lansing MI 48820	Ameritech	20.07	A13	ATI (= 288)	1MB	1
2	517–337–2445	517–337–2231	2001 Ross Avenue, E Lansing MI 48820	Ameritech	28.99	A13	ATI (= 288)	1MB	1
3	517–999–3499	517–999–3499	1717 Main St., E Lansing MI 48820	Ameritech	24.32	A13	MCI (222)	1MX	1
4	517–999–3499	517–999–1111	2001 Ross Avenue, E Lansing MI 48820	Ameritech	54.37	A13	ATI (= 288)	1ZR	1

Exhibit 46.   Sample Findings from the Business Line Worksheet

Category	Issue	Findings	Recommendation	Impact
Finding: Vacated location	Step 5 deliverable: Compare the service address with client-provided service addresses to locate addresses that have been vacated.			
Unnecessary circuits	Local lines	517–999–3499 was found to be working at a location vacated by client March 1999.	Place order with LEC to disconnect service. Advise IXC to remove number from database.	Take the monthly recurring charge, any usage charges, and the PIC charge to calculate the monthly savings.
Finding: Unused lines	Step 6 deliverable: Recommend service disconnection for unused lines.			
Unnecessary circuits	Local lines	No known purpose found for WTN 517–337–2225.	Disconnect service.	Take the monthly recurring charge, any usage charges, and the PIC charge to calculate the monthly savings.
Unnecessary circuits	ISDN	ISDN service 517–999–1111 no longer required.	Disconnect ISDN service.	Take the monthly recurring charge, any usage charges, and the PIC charge to calculate the monthly savings.
Finding: Unnecessary service	Step 7 deliverable: Compare monthly recurring costs to determine if lines are using unnecessary services.			
Unnecessary service	Unnecessary LEC service for inside wiring.	WTN 517–337–2231, used as a modem line, has a monthly charge for "inside wiring."	Advise LEC to cease billing for service.	Cost of service.
Finding: Unnecessary LPIC	Step 8 deliverable. Determine whether the LPIC can be changed.			
Least cost service	LPIC	The LPIC is not the preferred carrier for intraLATA calls.	Advise LEC to change LPIC.	Savings can be calculated by taking the preferred carrie′rs cpm and subtracting the current cpm.

Finding: Too many lines

Step 9 deliverable: Compare the number of lines to the number of people on location.

Unnecessary circuits	Local lines	The quantity of lines exceeds the recommended quantity.	Disconnect a line.	Take the monthly recurring charge and the PIC charge to calculate the monthly savings.

Finding: Premium charges

Deliverables template entry for step 2: Look at high cost services.

| Nonessential charges | Premium local charges | There were three charges for "Emergency Interrupt." | Educate staff about the cost of premium services, such as Call Return, Call Verify, and Call Interrupt. | $7.50 in monthly savings. |

Finding: Directory assistance

Deliverables template entry for step 3: Look at the cost of directory assistance.

| Nonessential charges | Premium local charges | There were 133 calls to local directory assistance at $.40 each. | Make available to staff local telephone directories and encourage the use of the printed directory. | $53.40 in monthly savings |

Finding: Nonessential charges

Deliverables template entry for step 1: Look for "slamming."

| Nonessential charges | Calls routed over wrong IXC on LEC bill | Line installed with PIC for carrier other than the preferred carrier. | Place order with LEC to change PIC. Advise IXC to put number on proper plan. | Calculate the cost difference in the plan rate and the billed charges. |

OR

| Nonessential charges | Calls routed over wrong IXC plan on LEC bill | Line installed with correct PIC, but IX not advised to put new service on plan. | Advise IXC to put number on the proper plan. | Calculate the cost difference in the plan rate and the billed charges. |

Exhibit 46.   Sample Findings from the Business Line Worksheet (Continued)

Category	Issue	Findings	Recommendation	Impact
Finding: Nonessential monthly charges	Deliverables template entry for step 2: Look at questionable monthly charges.			
Nonessential charges:	Monthly IXC charges on LEC bill	There was a $5 charge for a monthly plan.	Cancel service type that generates the $5 charge on LEC bill.	$5.00 in monthly savings.
Finding: Nonessential monthly charges	Deliverables template entry for step 3: Look for "cramming."			
Nonessential charges	Monthly IXC charges on LEC bill	There was a $25.14 charge for an item described as YP NET.	Cancel service if not needed.	$5.00 in monthly savings.

Exhibit 47.   Sample LEC Charges Worksheet

Item	BTN	WTN	Amount	Duration or Quantity	Call Type
1	214–953–0669	214–953–0669	5.30	00:37:48	IntraLATA toll
2	214–953–0669	214–953–0669	7.50	3	Emergency interrupt
3	214–953–0669	214–953–0669	53.40	133	Local information
4	214–953–0669	214–953–0669	23.25	06:36:00	Local measure service

Worksheet Analysis: LEC Charges

The LEC Bill – LEC Charges Worksheet (see Exhibit 47) should be reviewed for the type of calls, the resulting charges and the cost per call minute of certain types of calls.

Worksheet Analysis Steps

1. Calculate the cost per minute of intraLATA toll billed by the LEC. Compare with charges for intraLATA toll from the IXC. This example reflects a $.14 cpm for intraLATA calls. The $.14 should be compared to the IXC rate.
2. Charges for Premium Services such as "Emergency Interrupt" should be reviewed. The client should be made aware of the high cost for such services.
3. Charges for Premium Services such as "Local Information" should be reviewed.
4. Calculate the cost per minute of local measured service billed by the LEC. Compare with charge for like calls using the IXC. This example reflects a $.066 cpm for local measured calls. Compare the total cost of the measured service, line and usage, with flat-rate service if available. Compare the rates for like calls using an IXC. Using an IXC may require the dialing of additional digits, but most phone systems can be programmed to insert additional digits.

Worksheet Analysis: IXC Charges

The LEC Bill – IXC Charges Worksheet is used to review of the type of charges and the total spend per line type.

Sample IXC Charges worksheet

Exhibit 48 contains a sample that is an accumulation of charges from the LEC Bill – IXC Charges section.

Exhibit 48. Accumulation of Charges from the LEC Bill–IXC Charges Section

Item	LEC	IXC	BTN	WTN	Amount	Duration	Call Type or Charge Description
1	SWBT	AT&T	214–419–2900	214–419–2345	4.51	7	Collect
2	SWBT	American Tel	214–999–1400	214–999–1234	65.19	53	Dialed direct
3	SWBT	American Tel	214–999–1400	214–999–1234	5.00	NA	Monthly charge
4	SWBT	BPSI for YP NET	214–999–1400	214–987–2882	25.14	NA	Monthly charge

Worksheet Analysis

It is not unusual to find IXC charges from the carrier of choice. Collect charges are acceptable charges from an IXC on a LEC bill. AT&T may also bill collect charges to an AT&T Business Services Bill.

1. Charges for dialed direct calls billed by an IXC on an LEC bill may represent "slamming." However, there are several other issues to be reviewed before a final qualification of slamming is determined. The cost per minute should be calculated. This cost, less the contracted cost per minute, will be used to calculate savings.
2. Monthly charges billed by an IXC on an LEC bill are questionable. Fees for dialing plans should be questioned in light of special contracts with carriers. An IXC charge on an LEC bill typically occurs as a result of ordering additional lines but not notifying the preferred carrier's representative. Simply advising the LEC that AT&T (or any other carrier) is the preferred carrier does not get the line provisioned to get the contracted rates. Also, if there are not enough IXC trunks to carry all the long-distance traffic, the PBX will sometimes send long-distance calls through LEC trunks in order to get the calls through.
3. Monthly charges billed by an IXC on a LEC bill are questionable. Investigation of the source of this charge starts with a call to the toll-free number as seen on the individual bill page. This is an example of "cramming." The explanation for this charge is the listing of a phone number.

3.2 Analyze Contracted Rates versus Actual Billed Rates

Validate that the carrier is performing according to contract.by comparing the contracted rates with the actual rates for items charged.

- *Using the tariff rates and contracted discount, calculate the anticipated cost per minute per call type:* Material Gathering, Tariff and Contract Documents
- *Using the paper invoices or the CD-ROM files, calculate the billed cost per minute per call type:* Principal IXC 1+ Use Worksheet
- Make comparison of rates by type.

Calculation

For representative purposes, the prior tariff documentation will be used to calculate the contracted intrastate switched 1+ rate based on a discount of 35 percent.

- The average call is five minutes in length and a distance of 100 miles.
- Calculate prediscounted cost of call: $129 + (4 + (4 \times .189)$
- Prediscounted cost is $1.053
- Calculate discounted cost of call: $1.053 * (1 - .35)$
- Discounted cost is $.685.
- Calculate the cost per minute: $.685/5$

Exhibit 49.   Slight Differences between Contracted and Billed Rates

Type	Origination	Termination	Contract	Billed
Interstate	Dedicated	Switched	.08	.079
	Switched	Switched	.10	.014
	Dedicated	Dedicated	.05	.053
	Switched	Dedicated	.08	.079
	Toll-free number	Dedicated	.10	.099
	Toll-free number	Switched	.14	.1378
Intrastate:	Dedicated	Switched	.099	.101
State specific	Switched	Switched	.137	.138
	Dedicated	Dedicated	.07	.069
	Switched	Dedicated	.099	.101
	Toll-free number	Dedicated	.10	.098
	Toll-free number	Switched	.14	.1388

- Cost per minute is $.137.
- Perform function for each rate factor.

Table population

It is not uncommon to find slight differences between contracted and billed rates. This usually occurs because of call timing. Anything less than about $.005 may not be significant (see Exhibit 49).

3.3 Examine Trunk Reports

Trunks are flagged with a status code in the database. This code is determined after analyzing the trunk reports and the information in the database.

Trunk status

The following describes the different trunk statuses:

- *I:* in use
- *NI:* not in service (or in use)
- *NT:* not terminated in switch
- *INT:* in use and not terminated in switch
- *O:* out of service

Trunk Analysis: Determining Status for Avaya Switches

Before you begin

Complete the Request Switch Reports for Avaya Switches and Enter Switch Reports for Avaya Switches procedures before beginning this procedure.

Materials needed

Obtain the following reports from the Switch Administrator:

- List Trunk Group
- Display Trunk-Group Report <trunk group number>
- List Measurements Trunk-Group Hourly <trunk group number>
- List Measurements Trunk-Group Summary Today
- List Measurements Trunk-Group Summary Last
- List Measurements Trunk-Group Summary Yesterday
- List Performance Trunk Group Today
- List Performance Trunk Group Yesterday

Procedure

Follow these steps to determine the trunk status.

1. View the List Trunk-Group Report for each site.
2. Display assigned Trunk Group Number translated.
3. View the List Measurements Trunk-Group Hourly Report.
   a. If all columns contain zeros as values, go to Step 4.
   b. If not, go to Step 5.
4. Write Disconnect next to the Trunk-Group Number on the List Trunk Group Report.
5. View the Display Trunk-Group Report for each trunk group.
6. Obtain a Display Trunk Group Report (Take all trunk group numbers printed out from " List Trunk Group" and create a List measurements report. For example, Trunk Group Hourly and Trunk Group Numbers. This information is needed for any site trunked to the node if trunks between facilities are used.)
7. Enter the direction.
8. Match each Trunk Group Number using the trunk group quantity and service address with the corresponding WTN or circuit number in the WTN table.

3.4 Analyze Toll-Free Usage

Procedure

Follow these steps to look at locations where T1s are not installed.

1. Open the Toll Free.xls spreadsheet for a selected site.
2. Sort the **Amount** column by ascending order.
3. Are there any **Amounts** with less than $20 of usage?
   a. If yes, go to Step 4.
   b. If no, go to Step 11.
4. Copy the toll-free numbers from the **Toll Free No.** column.

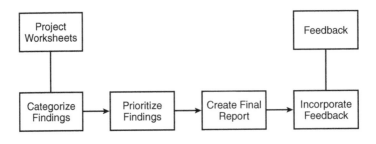

Exhibit 50.   Creating the Deliverables

5. Paste the numbers into a new document.
6. Copy and paste the corresponding usage charge from the **Amount** column into the new document.
7. Copy and paste the data in the **Purpose** column into the new document.
8. Provide the list of toll-free numbers to the client.
9. Does the client want to disconnect a number?
    a. If yes, perform Step 10.
    b. If no, ignore Step 10.
10. Enter *Disconnect* in the **Toll Free Status** column for the toll-free number.

4. Creating the Deliverables

Here, the individual savings items are documented and qualified for the final report.

Flowchart

The flowchart in Exhibit 50 shows the steps involved in this phase and the external entities that provide relevant information.

Create Deliverables

The following major steps are involved in the Create Deliverables phase.

4.1 *Categorize findings.* Gather all worksheet findings into a single matrix that will constitute the final report detail.
4.2 *Evaluate and prioritize findings.* Determine which findings are the most important and which provide the greatest savings. These are placed at the top.
4.3 *Create final report.* The project process, key findings, and recommendations are outlined in this report.
4.4 *Obtain and incorporate feedback:* Stakeholders provide feedback. Include feedback in the final report.

4.1 Categorize Findings by Issue

Organize all findings into a single matrix (Exhibit 51) by findings category.

4.2 Evaluate and Prioritize Findings

Determine which findings are the most important and which provide the greatest savings. These have the highest priority and are place at the top.

4.3 Create Final Report

The executive summary should take no more than two pages. All issue details should be included at the end as appendices.

Section Headings

- Executive summary
- Processes reviewed
- Company situation at time of project initiation
- Savings/recovery totals and breakout by location or region
- Source of savings/recovery and listing of priority issues
- Project scope and objectives
- Procedures
- Locations
- Individuals interviewed
- Results
- Issues and recommendations
- Detail should be in the appendices
- Cost savings issues (include pie chart with totals and percentages)
- General billing issues
- Recommendations
- Appendices

4.4 Obtain and Incorporate Feedback

Get All Questions Answered

Provide key company contacts with a draft copy of the finalreport. After discussion, include any additional information obtained in the final report.

Conclusion

After this generic work program has been tailored to your organization's architecture and business environment, it will significantly speed subsequent reviews. Generally, audits should be performed at least once every three years,

Exhibit 51. Findings Matrix

Category	Issue	Findings	Recommendation	Impact
Unused service	Local line Local trunks Circuits Toll-free service ISDN service	No known purpose Failed to disconnect when location vacated	Disconnect service, noting carrier, and other identifications, such as circuit number	Take monthly recurring charge, any usage charges, and PIC charges to calculate monthly savings
Switched versus dedicated	Outbound 1+ Inbound on-net 1+ Inbound on-net from calling cards	Incorrect provisioning Incorrect use of service Incorrect PBX programming	Change service	Calculate charges on recommended service and subtract actual charges for savings
Unauthorized service or charges	Slamming Cramming Wiring charges	Inappropriate or incorrect charges	Cancel service, noting service and provider	Savings are 100 percent of cost of item
Least cost service	Directory assistance Calling cards IXC usage LEC usage	Type of service described in greater detail	Change service with details of proposed replacement	Cost of service today, less proposed cost calculation, to produce savings
Unused service	Toll-free service	800-number from IXC has no usage but there is a recurring charge of dollars being billed	Cancel service	Take monthly recurring charge, any usage charges, and PIC charge to calculate monthly savings.
Unused service	ISDN	BellSouth billed this office $88.76 for ISDN service; most likely, this service was installed on a temporary basis prior to the installation of the AT&T Frame Relay	Disconnect the ISDN service	Take monthly recurring charge, any usage charges, and PIC charge to calculate monthly savings

Note: Further examples are contained in the analysis sections.

because technologies and organizations change enough that errors/waste can grow during the interim. Finally, at the risk of stating the obvious, routine audits, although valuable, are not a substitute for strong processes, internal control reviews, follow-up procedures, routine exceptions reports, and charge-back to the groups that incurred the expenses.

About the Authors

Brian DiMarsico is a senior manager in the Operational Effectiveness Group of PricewaterhouseCoopers responsible for Telecom and Network services in the New York Metro region. He has over thirty years of business and IT experience. Brian's experience includes a strong IT background in program management, application development, network, and business analysis within client server and mainframe computing environments. He has conducted many successful implementations of diversified computing technologies for the communication, insurance, banking, government, health, and manufacturing industries for users, clients, and product suppliers.

Brian is a member of the International Engineering Consortium and holds an MBA in Personnel/Industrial Relations, Fairleigh Dickinson University.

Thomas Phelps IV, CISA, is a manager in the Los Angeles Technology Security Practice of PricewaterhouseCoopers. He is the West regional lead for telecommunications security. Thomas has over 13 years of experience in information security, telecommunications, systems engineering, systems administration, semiconductors, two-way radio manufacturing, training, and journalism. He has worked with local, long distance, and wireless carriers in performing telecom controls reviews, including security audits of Mobile Switching Centers (MSCs). Thomas' security experience includes helping clients assess their enterprise security architecture, writing technical standards and controls, and managing security reviews for platforms such as routers, firewalls, UNIX, NT, Netware, PBXs and voice mail systems.

Thomas is the Vice President of the Los Angeles Chapter of the Information Systems Audit and Control Association (ISACA). He has authored several white papers and is a frequent speaker at IT audit and telecom events, including the North America Conference on Computer Audit, Control, and Security (North America CACS). He has contributed to a PricewaterhouseCoopers Customer Relationship Management book for the Information Systems Audit and Control Foundation (ISACF). With roots in Mt. Pleasant, Iowa, Thomas is a Beta Gamma

Sigma business graduate from the University of Texas at Austin and a member of the Honor Society of Phi Kappa Phi.

William A. Yarberry, Jr., CPA, CISA, is a partner with Southwest Telecom Consulting (www.SouthwestTelecomConsulting.com). He was previously a senior manager with PricewaterhouseCoopers, responsible for Telecom and Network services in the Southwest region. William has over twenty-four years of experience in IT application development, internal audit management, outsourcing negotiations and administration, and telecommunications management. His professional experience is in enterprise communications, with an emphasis on cost containment and server based voice technologies, such as CTI, IVR, and voice over IP.

He is the author of *Computer Telephony Integration* (published by Auerbach, 1999) and is currently writing a second edition, due to be published later this year. Prior to joining PricewaterhouseCoopers, William was director of telephony services for Enron Corp. He was responsible for the operations, planning, and architectural design for PBXs and related voice communications systems supporting over 7,000 employees. Now living in Houston, Texas, William is a University of Tennessee Phi Beta Kappa graduate in Chemistry and earned an MBA at the University of Memphis. He can be reached at billyarberry@bigfoot.com

Index